WI-FI ENABLED HEALTHCARE

WI-FI ENABLED HEALTHCARE

Ali Youssef • Douglas McDonald II
Jon Linton • Bob Zemke • Aaron Earle

 CRC Press
Taylor & Francis Group
Boca Raton   London   New York

CRC Press is an imprint of the
Taylor & Francis Group, an **informa** business
AN AUERBACH BOOK

CRC Press
Taylor & Francis Group
6000 Broken Sound Parkway NW, Suite 300
Boca Raton, FL 33487-2742

Printed on acid-free paper
Version Date: 20140121

International Standard Book Number-13: 978-1-4665-6040-6 (Hardback)

Library of Congress Cataloging-in-Publication Data

Youssef, Ali.
    Wi-Fi enabled healthcare / Ali Youssef [and four others].
        pages cm
    "A CRC title."
    Includes bibliographical references and index.
    ISBN 978-1-4665-6040-6 (hardcover : alk. paper)
    1. Medical telematics. 2. Medical care--Data processing. 3. Wireless communication systems. I. Title.

    R119.95.Y68 2014
    610.285--dc23   2013045729

Visit the Taylor & Francis Web site at
http://www.taylorandfrancis.com

and the CRC Press Web site at
http://www.crcpress.com

Contents

Foreword

Rapid advancements in wireless technologies are transforming how healthcare is delivered, extending care and access to critical health data anywhere, anytime. This transformation presents health systems and care providers with a host of opportunities and challenges inside and outside their facility walls. The unprecedented speed with which these wireless and telecommunications advancements have converged upon health systems has led to an urgent need for information technology, biomedical, and telecommunication professionals to understand wireless architectures and the technical, regulatory, fiscal, and policy implications for implementing wireless networks in healthcare today and tomorrow. As wireless technology and processing speeds continue to evolve, healthcare providers can expect the demand for and use of more sophisticated untethered care solutions to increase. A focus on infrastructure to provide a solid, safe, secure foundation for these new care solutions is critical. This book seeks to close the knowledge gap on wireless infrastructure and provide practical technical guidance for health systems providers to ensure their systems provide reliable, end-to-end communications necessary to surmount today's challenges and capitalize on new opportunities as this technology evolves.

Highlights of wireless opportunities for healthcare providers include improvements in

- Workflow: point-of-care delivery and workflow enhancements provide remote and bedside registration, diagnostics, and treatment, as well as staff and patient tracking.
- Communications: real-time connectivity between nurse, staff, and patients.
- Transportation: real-time connectivity to emergency medical services and transport services, allowing for the transfer of critical information while patients are in route between care settings or departments, or in the home.
- Consumer engagement: consumers and care providers may now interact through remote communications and monitoring devices, enabling clinicians and patients to communicate timely health information, reminders, and support to each other in real time, changing patient–caregiver relationships.
- Workforce shortages: provides infrastructure for new care models and a flexible mobile workforce.
- Asset management: provides new tools for asset tracking.
- Data access: allows for the ability to collect, analyze, and share critical patient data, including access to electronic health records and health information exchange.
- Usability: provides introduction to consumer-based devices with a high level of user-centered design, improving ergonomics, and user interface flexibility.
- Innovation: provides the foundation for new applications such as Body Area Networks, deploying body sensors, untethering patients from monitoring devices, diagnostic testing equipment, and the need to remain in traditional health facilities for observation and treatment.

Challenges of wireless technologies include:

- Privacy and security: ensuring data and patient confidentiality are secure through both technical means and operational policies is essential.
- Regulatory requirements: federal, state, local, and institutional regulations may be nonexistent and/or may vary with regard to definitions of mobile medical device applications, physician and provider licensure and liability for use, etc., effecting how these tools are to be deployed and used.

- Infrastructure coexistence: very few healthcare providers have the luxury of building wireless infrastructure from scratch. A multitude of applications exist inside facilities, such as wireless LAN, telemetry, cellular and public Wi-Fi, with hundreds if not thousands of untethered devices producing interference and security challenges. Lead walls, elevator shafts, and historical piecemeal construction challenge essential reliable coverage.
- New infrastructure: staying abreast and understanding the technical, policy, and procedural requirements of new policies such as mBAN spectrum capacity and allocation is essential, but can be daunting.

Surpassing these challenges and capitalizing on current and future opportunities will require a solid understanding of wireless infrastructure. The shared experience and lessons learned from the authors provide essential guidance for large and small healthcare organizations in the United States and globally.

Edna Boone
Office of National Coordinator of Health (ONC)

Preface

Why write a book focused on wireless in healthcare? If you are interested in this topic chances are it's because you are somehow involved in this space either from IT operations, IT leadership, clinical engineering, healthcare administration, or a related field.

The backgrounds of the authors vary from network engineering to IT security, to biomedical engineering. Our knowledge is founded upon formal study and graduate studies, but what we have to offer that is unique comes from many hours spent in the trenches of healthcare IT operations. What we all have in common is that as we began designing, deploying, and supporting wireless networks for various healthcare accounts, we soon learned that these types of inpatient and outpatient facilities have unique mobility requirements that lead to interesting challenges. During the early years of WLAN deployments at the turn of the twenty-first century, most organizations that we jokingly referred to as "cube lands" had relatively simple requirements of employee laptop connectivity in conference rooms and workspaces. Seamless roaming, handheld devices, guest access, and mobile medical devices were years away from becoming mainstream. We were fortunate to be working in a complex environment that from the beginning had greater demand for mobility, complex user requirements, unique radio frequency challenges, and a plethora of use cases for mobile devices. Whitepapers on best practices for design and support did not seem to cover the areas that we were working to address, such as clinicians with personal devices (including access points), VoWLAN coverage in elevators, and FDA-certified biomedical devices. BYOD was not a term a decade ago but that did not stop the demands for employee and patient personal devices on the networks.

What was out there was vendor-specific marketing focused around how their technology could solve all of our mobility aspirations. Sounds familiar? As our projects grew in scope, complexity, and outright quirkiness we began to document operational run books for the teams. Technology choices are only a small component of the operational support challenges that await a network deployment. These ops manuals become the basis for our architecture standards and best practices guidelines for support. Lessons learned in the trenches so to speak. As the wireless standards evolved from 802.11b to 802.11n, and mobile devices grew from a handful of Microsoft PDAs to thousands of IOS clients, so have our ops manuals. The one constant we have seen is that dependency and mission criticality of the wireless network is growing with no signs of slowing down. With this in mind the team thought we would share our experiences and lessons learned, and provide a guide that we could have made use of when we first embarked on our wireless journey in one of the largest healthcare systems in the country. We hope it will be of help.

1

BRIEF HISTORY OF WI-FI

The concept of transferring data over electromagnetic radio waves has been a topic of interest dating back to the 1800s. Although Vic Hayes is considered by many to be the "father of Wi-Fi," the wireless standards he helped craft rely on fundamental principles that are rooted in the work of intellectuals in the field of electromagnetism like Heinrich Hertz, James Clerk Maxwell, and Michael Faraday. Other inventors such as Guglielmo Marconi, Nikola Tesla, Alexander Popov, Sir Oliver Lodge, Reginald Fessenden, Amos Dolbear, Mahlon Loomis, and Nathan Stubblefield have also helped shape our modern wireless communications. Out of all of these inventors, the ones that stand out the most are James Clerk Maxwell and Heinrich Hertz. Maxwell in 1864 crafted a set of mathematical equations predicting the existence of radio waves and likening them to light waves. Between 1885 and 1889 Heinrich Hertz (Figure 1.1) was able to produce electromagnetic waves in a series of controlled laboratory experiments, making him the first scientist to successfully send and receive radio waves. He was also able to measure their wavelength and velocity, which helped bring validity to the Maxwell equations. Hertz documented his work in the *Annalen de Physic und Chemie*, which later served as the foundation for the design and development of radio and electromagnetic wave transmission systems.

Each of the scientists and inventors mentioned above relied on and referenced the work of the others and built upon it. The ongoing evolution of this field is a great showcase of how modern inventors and scientists continuously refine and improve each other's work. This is a great example of the origin of the phrase "we stand on the shoulders of giants."

Although some of the inventors referenced above had successfully transmitted electromagnetic radio frequency (RF) signals over several kilometers, the technology did not experience its first major

Figure 1.1   Heinrich Hertz. Source: Wikipedia Public Domain.

boom until the 1940s, especially around the time of World War II. The U.S. military had a keen interest in using the technology to disrupt enemy communications and secure their internal communications. With the help of a mathematician, who also happened to be one of the first actresses to appear nude on camera, Hedy Lamarr (Figure 1.2), techniques for frequency hopping and spread spectrum communications were born in 1942.[1] The anti-jamming capabilities of

Figure 1.2   Hedy Lamarr. Source: Wikipedia Public Domain.

this electromagnetic RF technique drove the military to invest heavily into developing the technology. Spread spectrum kept evolving and becoming more sophisticated through the 1970s right up to the beginning of the 1990s.

Hawaii serves not only as a wonderful vacation location with its pristine beaches and blue skies, it was a major driving force in revolutionizing the wireless communications field in the 1970s. It is widely agreed that the first wireless network used for data transmission in the public sector was one developed at the University of Hawaii in 1971. This network, known as ALOHAnet, was designed to connect seven computers on four Hawaiian Islands wirelessly to a central computer located on the Oahu campus. The network was one of the first to utilize a shared medium for the wireless clients, which inevitably led to packet collisions. The technology researchers at the university took several steps to get around the issue. Initially, they tried to assign dedicated time windows when each client was allowed to transmit its packets, regardless of whether the client had packets to transmit. The inefficiency of this approach gave rise to carrier sense multiple access (CSMA), where each client could listen in on the channel to determine if it was in use before transmitting its data. To prevent a scenario where multiple clients could transmit their data at the same time, the inventors added the ability of the client to determine if its packets had made it successfully to the central hub. This ultimately resulted in CSMA-CD or carrier sense multiple access–collision detection. This innovation later became a foundation in the development of Ethernet communications as well as Wi-Fi.[2]

History and Current Growth and Proliferation of Wi-Fi in Hospitals

Here in the United States, wireless communications did not start propagating quickly within healthcare institutions until 1985 when the Federal Communications Commission (FCC) made unlicensed spread spectrum available in the ISM (Industrial, Scientific, and Medical) bands. Although second generation (2G) cellular technology was in use at the time, the bulky handset designs and the limited bandwidth along with high recurring monthly subscription costs made this option less attractive. Following the FCC decision, manufacturers started developing different types of mobile devices that operate in the unlicensed frequency ranges. The lack of standardization caused confusion and made interoperability challenging. Some of the earliest adoptions of this technology in healthcare were wireless phones that operated in the 900-MHz range. These were ideal for clinicians and nurses to be accessible while being mobile throughout clinics and hospitals. The handsets required dedicated 900-MHz base stations installed throughout the facility. Although having a dedicated frequency for the voice devices decreased the possibility for medium contention, it was costly. To date, many hospital networks still have a dedicated overlay 900-MHz system and are in the process of evaluating potential replacements.

A major breakthrough occurred in 1997 when the original IEEE 802.11 protocol was ratified. Although the supported data rates were only 1 to 2 Mbit/sec, the promise of interoperability was attractive to healthcare institutions, and Wi-Fi capable barcode scanners used for inventory management began to be used more heavily in hospitals. The 802.11 standard also defines a specification for an infrared (IR) physical layer, which was primarily intended for ad hoc, short-range networking. This technology was primarily used for medical device diagnosis and troubleshooting, but was never adopted as a mainstream gateway for allowing devices onto hospital networks.

The two unlicensed frequencies that saw the highest utilization were the 2.4-GHz and the 5-GHz bands. In light of this, the IEEE 802.11b and 802.11a amendments were released in 1999. Almost immediately, medical device manufacturers, hospital systems, and clinics began to take an increasing interest in adopting the technology. The quick adoption was very unusual for an otherwise very technologically conservative group of professionals and in many ways forced hospital IT departments to advance the technology faster than their peers in other industries. Potential workflow improvements due to mobility were clear, and physicians started to be drawn to the allure of Wi-Fi-capable devices. At the time, the most popular type of handheld device was the personal digital assistant (PDA) and clinicians began to leverage these types of devices to run simple applications like general medical reference, drug interactions, and active sync to synchronize their calendars and tasks. The early PDAs were devices like the Palm Tungstens, Sony Clie; and Pocket PC devices, which include the Compaq iPAQ. Between 1999 and 2006 there is clear evidence of an increasing trend in PDA use. The adoption rate for individual professional use ranged up to 85 percent in 2006 indicating a high rate of adoption among physicians.[3] Although the additional mobility was handy in areas like the intensive care unit (ICU), the limited processing capabilities and onboard memory of these types of devices left early adopters wanting more. The explosion in smart phone utilization is covered later in the chapter, but it is important to note that PDAs were the starting point.

Up until the 1990s, mainframe "green screens" were still being slowly replaced by desktop computers for administrative, billing, and clinical data collection repositories using 10- and 100-Mbps wired Ethernet interfaces. With the evolution of 802.11b, networking manufacturers began offering Personal Computer Memory Card International Association (PCMCIA) cards to retrofit existing laptops. Shortly after, laptop manufacturers began aggressively integrating wireless cards into their devices and fine tuning their designs for portability. At the same time many hospital IT departments began to discover how easy it was becoming for clinicians to install their own consumer home routers to provide connectivity. By 2001 Wi-Fi chipset shipments reached 7.5 million units.[4]

The additional available bandwidth allowed clinicians to use these types of devices to access the Internet, and check their e-mail at reasonable speeds. During this same period, attention was being paid to the cost of medical errors, and the need for computer order entry and rudimentary access to electronic medical records (EMRs) on mobile devices. It would take a decade for these applications and the hardware devices to run them to mature, and for healthcare institutions to have enough government-sponsored programs and incentives to aggressively pursue these types of systems. Mobility was, and still remains, a key driver.

In 2000, most medical devices did not have integrated wireless capabilities, but medical device engineers and manufacturers began accounting for the capability in their product designs. (This will be covered in more depth in Chapter 6.) Another key catalyst was the fast proliferation of the technology in homes around the world. This left many professionals including clinicians wanting to start using and taking advantage of the same technology at their workplace. 802.11b was much more popular than 802.11a due to its lower cost, wider

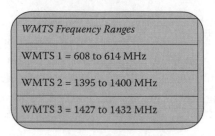

Figure 1.3   WMTS frequencies. Source: FCC website.

availability, and larger RF signal footprint. Although the data rates of 11 Mbps were sufficient for most medical devices, phones, and other wireless handheld devices, it was not enough to efficiently allow the transmission of radiology images or video over Wi-Fi. 802.11a allowed data rates up to 54 Mbps, but was not adopted nearly as quickly as 802.11b. This is mainly because upon its initial release 802.11a's network interface cards (NICs) cost 50 percent more and its access points were being priced 35 percent more than their 802.11b counterparts.

In 2000, the FCC dedicated a portion of the radio spectrum to wireless medical telemetry systems (WMTS). Figure 1.3 outlines the designated frequency ranges.

This was widely adopted for remote monitoring of a patient's health. WMTS is used to transmit vital signs from a given medical device to a remote nurses' station equipped with a radio receiver.[5] Although the technology worked well and was effective, it left a void because it did not allow use of voice or video applications. This technology was reserved for healthcare facilities and providers that offer services that last longer than 24 hours. Almost all hospitals today still utilize some form of a WMTS. These types of systems work, but they suffer from channel interference and mostly rely on unidirectional communication with no packet loss or retry mechanism. In addition to WMTS, medical device manufacturers leverage Bluetooth and Zigbee for personal area network communication. These technologies are well suited for ad hoc communication between a medical device and its associated sensors. In 2012, the Middle Class Tax Relief and Job Creation Act directed the FCC to relocate channel 37 and provided instructions on how to compensate WMTS licensees. This will impact telemetry systems occupying the 608- to 614-MHz band. This could be an early sign that telemetry equipment vendors will try looking at leveraging Wi-Fi for transmitting patient vitals. This approach would decrease the total cost of ownership significantly due to leveraging existing Wi-Fi networks and utilizing security mechanisms like 802.1x that decrease liability.

In 2003, 802.11g was ratified by the IEEE. It worked on the same frequency and was backwards compatible with 802.11b, but it allowed for data rates up to 54 Mbps. 802.11g utilized a similar modulation as 802.11a. This resulted in the first major boom of 802.11-based wireless devices in healthcare. By 2005 Wi-Fi Chipset shipments had topped 100 million annually.

With the pent-up demand for additional wireless bandwidth, a plethora of new wireless devices were added to the wireless network. For the first time, it seemed feasible to run video or voice applications over the wireless network. Workstations on wheels, or CoWs as they were initially called, became very common in healthcare, and there was an increase in mobile cart vendors. In addition, for the first time, it was feasible to provide guests with Internet access without having to be overly concerned about bandwidth contention with the limited 11 Mbps. With the release of 802.11g, a greater focus was placed on leveraging mobile technologies for preventing medical errors, computerized order entry, and disease management. Major EMR vendors like EPIC Systems and Cerner were growing aggressively and had already developed a robust portfolio of EMR platforms. These could now start to leverage mobile devices.

802.11g was sufficient for most applications, but the evolution of mobile applications and the increasing number of devices that were being developed to leverage the wireless network and the data, video, and voice convergence revolution required higher data rates. 802.11n was released in 2009 allowing for data rates up to 600 Mbps and allowing for backwards compatibility with 802.11a,b/g. To date, 802.11n has been a game changer. With speeds greater than traditional wired speeds, market penetration has been impressive. Mobile device manufacturers have embraced 802.11n, and it has been forecasted that 1.5 billion products equipped with 11n will be sold in 2015, more than double the estimated 700 million in 2011.

Medical device manufacturers have also been slow to integrate 802.11n wireless network cards due to the backwards compatibility with 802.11 a/b/g. Most medical devices in hospitals currently support 802.11 b/g, with the more carefully designed devices supporting

Protocol	Date of Release	Freq. (GHz)	Modulation	Healthcare Adoption
802.11	Jun 1997	2.4	DSSS, FHSS	Slow
802.11a	Sep 1999	5	OFDM	Slow
802.11b	Sep 1999	2.4	DSSS	Moderate
802.11g	Jun 2003	2.4	OFDM, DSSS	Aggressive
802.11n	Jun 2009	2.4/5	OFDM	Aggressive

Figure 1.4   802.11 a/b/g/n adoption.

the less congested 802.11a band. Only recently did wireless voice handset vendors start integrating 802.11n chipsets into their products. The 802.11n clients can continue to use the 802.11 a/b/g infrastructure (Figure 1.4).

With 802.11n wireless access point sales comprising 75 percent of all access point sales, a growing number of hospitals have upgraded their wireless infrastructure to 802.11n, but medical device vendors are still in the process of ramping up. With the additional available bandwidth, hospitals are using the wireless network to:

- Offer guest services.
- Leverage real-time location services.
- Provide video remote interpretation.
- Support telemedicine.
- Offer voice over Wi-Fi integration with the nurse call/alerting system.
- Accommodate wireless medical devices.
- Provide access to EMRs and other healthcare applications.

"Since the birth of the iPhone and the Android operating system, mobile data use has exploded. By 2015 traffic will be some 20 times its 2010 level."[6] Offloading traffic onto the wireless network from the cellular network will be required as the cellular spectrum becomes more utilized. The release of the Apple iPhone, iPad, and some of the other tablets and phones available from competitors, has accelerated the "bring your own device—BYOD—to work" phenomenon, which is slowly becoming more popular and one more technical hurdle to be overcome on the wireless network in hospitals. It has also created an interesting paradox of using the cellular network for data when outside of the hospital while wanting to leverage the Wi-Fi network for voice when inside the hospital.

As the Wi-Fi technology has evolved, so has cellular technology. As seamless roaming between the two systems is becoming a reality, the ramifications for clinical workflows and healthcare in general are growing. These technologies could be used to drive healthcare costs down and facilitate remote patient diagnosis and treatment. For more developed countries this would mean streamlining patient care and allowing clinicians to view critical clinical patient data in real time. This would help expedite care to patients living in rural communities with no major hospitals in the vicinity.

Many hospital systems are deploying Wi-Fi systems with the hope of long-term cost savings by maximizing the types of applications and devices that utilize the wireless network. One of the key drivers is the demand for wall-to-wall mobility using the Wi-Fi network throughout health systems. This is to allow clinicians and other staff members to access the voice network anywhere within their facility without having to rely on cellular repeaters or distributed antenna systems. Providing adequate coverage in certain areas of the hospital can be challenging. This will be discussed further in Chapter 3 on the site survey process.

With the growing need for cost savings and the competitive nature of healthcare institutions, nurses and clinicians need to have immediate access to patient data at the bedside for information such as X-ray or blood gas results and access to the EMR system. This allows them to make informed decisions quickly while they are near the patient and to spend more time communicating with the patient. This is extremely important given the trend of some hospitals trying to mimic the hotel experience. If two hospital systems have equivalent doctors, a key differentiator can be patient satisfaction.

As the concept of unified communications continues to grow and spread the message of a work-life balance, BYOD is becoming a reality in many healthcare systems. Devices that clinicians are interested in bringing into the hospital are mobile phones, tablets, and laptops. These devices rely on Wi-Fi connectivity. This, together with concerns for improving the guest/patient experience by providing complimentary Internet access, is a major driver for deploying Wi-Fi.

Cellular providers that have traditionally been opposed to leveraging Wi-Fi in fear of losing subscribers have become fans of the technology and are helping drive offloading wireless traffic onto Wi-Fi whenever possible. The FCC expects a spectrum deficit by the end of 2013. With the cellular frequencies on the way to becoming saturated, cellular providers are opting to dump traffic onto local area networks whenever possible.

Although the healthcare industry has taken a prudent approach to the adoption of mobility, federal legislation is driving the need for accelerated adoption. Financial incentives from the federal government for attaining meaningful use and EMR implementation are helping drive this trend. With the introduction of the iPhone, iPad, and other smart phone and tablet platforms, access to EMR on mobile devices has become a priority. Healthcare IT strategies that are focused beyond Stage 3 incentive payments and having a comprehensive EMR will be critical to ensuring that institutions are providing cutting edge, competitive services. Wireless devices are a critical piece of this trend, and continue to play a vital role in improving the quality of care, patient safety, and reducing the overall cost of healthcare.

Many healthcare institutions are now looking at leveraging Wi-Fi for Real-Time Location Services (RTLS). This includes inventory tracking and management of biomedical and other types of devices, hand-washing tracking initiatives, security distress badges, and remote temperature monitoring. Several departments such as clinical engineering, security, and pharmacy are able to utilize the same wireless network to meet their departmental needs. The return on investment of such systems can be significant, and will be covered in depth in Chapter 9.

802.11n has also resulted in an increased appetite for voice over Wi-Fi and video over Wi-Fi applications. The use cases range from remote video interpretation for the hard of hearing to full-blown mHealth applications. Some other use cases are video conferencing, patient and employee training, virtual doctor visits, video consultation, dictation, and patient monitoring.

Regulatory Bodies

This section describes organizations that are involved in regulating Wi-Fi technologies, but not necessarily medical devices or other types of equipment used in hospitals. The focus is on organizations such as the FCC, IEEE, Wi-Fi Alliance, and ISO, but not AAMI, or the

Food and Drug Administration (FDA). These other organizations will be covered in depth in Chapter 4.

Federal Communications Commission

The Federal Communications Commission (FCC; www.fcc.gov), established by the communications act in 1934, is responsible for regulating all domestic and international radio, satellite, cable, wire, and television communications. This organization is an independent branch of government that has oversight from the U.S. Congress. Although they have been in some high-profile cases regarding regulating public radio transmissions, the FCC helped mold the evolution of Wi-Fi by making the unlicensed spread spectrum available in the ISM bands. Traditionally the FCC sets rules for RF communications without consulting other organizations, but lately there has been a push to communicate more closely with the FDA. This is to ensure that there is sufficient oversight for wireless medical devices and facilitating mHealth, the practice of medicine using mobile devices. A great example of this collaboration is the spectrum that the FCC reserved for the Medical Personal Area Network in 2012. This provides medical device manufacturers with an alternative to using the congested unlicensed wireless spectrums for wireless communications between medical sensors and medical devices. The FCC maintains a rich website with information about all of its licensed as well as unlicensed wireless frequency allocations. With the various wireless systems in hospitals, it is not unusual to have to refer to their site to identify what a specific frequency is being utilized for. Wi-Fi is specifically covered in Title 47: Telecommunications, Heading 15: Radio Frequency Devices. The FCC addresses details like radio power limits, antenna, frequency, and labeling requirements.

Institute of Electrical and Electronics Engineers

The Institute of Electrical and Electronics Engineers (IEEE; www.ieee.org) is the world's largest technical professional association dating back to 1884. There are currently over 407,000 engineers, scientists, IT professionals, physicists, doctors, and other professionals that are members of the organization. The IEEE publishes standards on a wide range of electrical systems ranging from commercial power systems to harmonic power controls. They also address standards for medical device communications at a high level. From a wireless perspective, the IEEE develops and documents wireless telecommunications standards. Wi-Fi is addressed in depth under the 802.11 project. The work that the IEEE has done on the 802.11 a/b/g/n standards is relevant to the consumer as well as medical devices. They have also worked extensively on the alphabet soup of standards, exhausting the entire alphabet and circling back around and using two letters to identify a standard (for example, 802.11ac and 802.11ad). Many of these have been and will continue to play a critical role in improving the Wi-Fi experience in healthcare. The full list can be reviewed in the appendix, but several of the standards have had and will continue to have a significant impact in healthcare. These are 802.11i, 802.11e, 802.11r, 802.11w, 802.11h, 802.11k, 802.11v, and 802.11u. Their significance is outlined in the next few paragraphs.

With the exception of 802.11 a/b/g/n, 802.11i is perhaps one of the most influential IEEE standards in healthcare. This standard defines the type of authentication and encryption that must be used to establish a Robust Security Network (RSN). In a nutshell, this is to utilize port-based authentication (802.1X) in conjunction with Advanced Encryption Standard (AES) encryption. The detailed mechanisms are covered in the security chapter of this book. This is relevant to healthcare in that 802.11i facilitates meeting regulatory requirements like Health Insurance Portability and Accountability Act (HIPAA) and Health Information Technology for Economic and Clinical Health (HITECH), ensuring that patient data is protected. Several large medical device vendors, including GE, Phillips, and Carefusion have taken the initiative and made their products 802.11i compliant, but these are the exception, not the rule. Larger healthcare institutions often use 802.11i as a benchmark for allowing wireless medical devices onto their networks. This is to try and avoid security vulnerabilities that can be the result of having clients using legacy forms of encryption on the network.

802.11e defines quality of service enhancement over 802.11 wireless networks. This is critical in healthcare due to Wi-Fi being a shared medium with limited bandwidth. Quality of Service (QoS) is essential to ensure that time-sensitive applications like video over wireless (telemedicine) or voice over wireless systems are guaranteed a certain amount of priority on the network. In addition QoS mechanisms allow clinical diagnostic traffic to have a higher priority than unessential communication. This is to ensure consistent, repeatable, and adequate wire-like performance over Wi-Fi.

It is fairly common knowledge that the roaming burden on a wireless network relies on the roaming algorithm built into the specific Wi-Fi client in use. The client makes the decision to roam or to remain associated with a given wireless access point based on its algorithm. 802.11r addresses some key enhancements to improve client roaming by allowing for fast handoff between access points. One of the key areas impacted is the ability to seamlessly roam across access points while implementing 802.11i for voice. Roaming is often an Achilles heel within healthcare facilities. The combination of very densely deployed Wi-Fi networks (micro cells) and legacy technology on devices make this a key amendment. 802.11r does a fantastic job of addressing roaming for applications that are intolerant or sensitive to delays.

A main reason that IT departments steer clear of adding life-critical devices onto their Wi-Fi network is the fact that these networks are susceptible to denial of service attacks. One of the best known attacks that cannot be prevented to date is RF jamming. Whether intentional in nature or not, this is comprised of introducing a high-powered RF signal to disrupt the airspace where 802.11 systems are operating. There are other types of denial of service attacks that involve packet manipulation and injection that can be prevented. 802.11w does not prevent RF jamming attacks, but it adds certain layers of security to increase the security of management frames. This is important in a healthcare setting, because it adds layers of security to discourage attempts to launch denial of service attacks.

802.11h is an IEEE amendment that defines transmit power control (TPC) and dynamic frequency selection (DFS). This was dedicated to managing the 5-GHz band and ensuring that it does not interfere with radar transmissions. It also introduces more available frequency space as part of the UNII 2E band. This is useful in healthcare settings for flexibility with channel planning, and for ensuring that the 5-GHz systems are staying clear of frequencies used for radar transmissions.

With so much of the wireless performance relying on the wireless client, 802.11k provides a way for radio resource measurements, logging data points such as transmit power control, client traffic statistics and errors, RF channel statistics, and neighboring report. This standard significantly improves client roaming and performance, and also allows network administrators to quickly zoom in and find potential network configuration issues.

To facilitate large-scale wireless client deployments, 802.11v allows for centrally managed client configuration, client load balancing, and deploying security configurations in a streamlined manner. With the growing size of healthcare systems, it is not unusual to find systems with one hundred or more hospitals and clinics. 802.11v is part of what allows the use of centrally managed wireless controllers servicing multiple campuses. This is essential to keep management overhead costs low.

MHealth, a subset of Ehealth, has experienced explosive growth in the last few years. It entails clinicians being able to diagnose patients in remote locations using a variety of wireless networks, devices, and applications. Some of the applications run on mobile devices that are able to leverage cellular networks. Wi-Fi offloading is becoming critical to offload some of the high-bandwidth requirements of these types of applications. 802.11u, completed in 2011, is an amendment to the 802.11 standard which adds features to improve internetworking and seamless roaming between different types of wireless networks.

Wi-Fi Alliance

The Wi-Fi Alliance (www.wi-fi.org), established in 1999, is a non-profit association of companies dedicated to the vision of seamless connectivity and to promoting the adoption of Wi-Fi around the globe. Much of the content in this section is quoted directly from the Wi-Fi alliance website. The mission of the Alliance is to:

- Provide a highly effective collaboration forum.
- Grow the Wi-Fi industry.
- Lead industry growth with new technology specifications and programs.
- Support industry-agreed standards.
- Deliver great product connectivity through testing and certification.

In an effort to meet these goals, the Alliance established the Wi-Fi CERTIFIED™ program in 2000 to test products for compliance with the 802.11 industry standards for interoperability, security, easy installation, and reliability. The Wi-Fi CERTIFIED logo is verification that the Wi-Fi Alliance has tested a product for compatibility with other Wi-Fi CERTIFIED equipment that operates in the same frequency band. Designated test labs conduct interoperability testing to validate and document that wireless devices work together and support secure connections.

Core Programs
- Wi-Fi products based on IEEE radio standards: 802.11a, 802.11b, 802.11g in single or dual (802.11b and 802.11g) mode, or multiband (2.4- and 5-GHz) products.
- WPA2™ (Wi-Fi Protected Access 2): Wi-Fi wireless network security, which offers government-grade security mechanisms for personal and enterprise use.
- EAP (Extensible Authentication Protocol): an authentication mechanism used to validate the identity of network devices (for enterprise devices).
- Protected Management Frames: Wi-Fi CERTIFIED WPA2 with Protected Management Frames provides a WPA2 level of protection for unicast and multicast management action frames.
- Wi-Fi CERTIFIED n: the latest generation of Wi-Fi operation, which supports the IEEE 802.11n ratified standard. This test program also includes Wi-Fi Multimedia (WMM) testing.
- Wi-Fi CERTIFIED ac: the first generation of Wi-Fi that can deliver up to gigabit per second data rates. Based on IEEE 802.11 ac, this program requires devices to successfully pass all certified n tests.

Optional Programs
- Wi-Fi Direct™: a certification mark for Wi-Fi client devices that connect directly without use of an access point, to enable applications such as printing, content sharing, and display. Wi-Fi Direct certifies products that implement

technology defined in the Wi-Fi Alliance Peer-to-Peer Technical Specification.

- Wi-Fi Protected Setup™: facilitates easy set-up of security features using a Personal Identification Number (PIN) or other defined methods within the Wi-Fi device. Wi-Fi Protected Setup certifies products that implement technology defined in the Wi-Fi Simple Configuration Technical Specification.

- WMM® (Wi-Fi Multimedia™): support for multimedia content over Wi-Fi networks enabling Wi-Fi networks to prioritize traffic generated by different applications using QoS mechanisms. WMM certifies products that implement technology defined in the WMM Technical Specification.

- WMM-Power Save: power savings for multimedia content over Wi-Fi networks, which helps conserve battery life while using voice and multimedia applications by managing the time the device spends in sleep mode. (In recent testing, this showed 37 to 73 percent power savings versus legacy power save mechanisms.)

- Voice-Personal: voice over Wi-Fi, which extends beyond interoperability testing to test the performance of products and help ensure that they deliver good voice quality over the Wi-Fi link.

- CWG-RF: for converged devices with both Wi-Fi and cellular technology, which provides detailed information about the performance of the Wi-Fi radio in a converged handset, as well as how the cellular and Wi-Fi radios interact with one another. This is now mandatory for Wi-Fi enabled handsets seeking CTIA certification.

- Voice-Enterprise: certifies that products are able to meet requirements supporting good voice call quality and advanced WPA2™-Enterprise security mechanisms. The program supports fast transitions between access points and provides management for voice applications. Voice Enterprise builds on the Voice-Personal certification features.

- WMM-Admission Control: enhanced bandwidth management tools to optimize the delivery of voice and other traffic in Wi-Fi networks. WMM-Admission Control certifies products that implement technology defined in the WMM Technical Specification.

- IBSS and Wi-Fi protected setup: designed to ease setup of connection for devices with limited user interfaces.

The FCC has been working closely in partnership with Continua Health Alliance to promote the Wi-Fi certification program on wireless medical devices. The organization released several white-papers outlining Wi-Fi and Wi-Fi security in healthcare in 2011 and 2012 that provide details on best practices and promote their certification program.

International Organization for Standardization

The International Organization for Standardization (ISO) is a global nongovernment agency with the goal of enabling a consensus to be reached on solutions that meet both the requirements of business and the broader needs of society. It is comprised of representatives from 164 countries. ISO is responsible to for developing the Open Systems Interconnection (OSI) model, which is the foundation of all network communications. OSI is comprised of seven layers, as outlined in Chapter 7. Wi-Fi defines communications in the Physical and Data link layers of the OSI model.

Wi-Fi Impacts on Clinical Workflow

The introduction of Wi-Fi in healthcare has dramatically changed the traditional clinical workflow, and added some new, exciting opportunities. In the 1990s, it was common to see rows of filing systems at a typical physician's office, and medical records were a series of printouts in the folder with charts, lab information, health records, and so on. Today, we rarely run into this scenario. It is much more likely that your clinical data is housed in a computerized EMR system. This cannot be attributed to Wi-Fi, but is rather tied to mandates and direction from the federal government. Wi-Fi did improve one component, and that is the potential for more interaction with physicians as they review patient records. Instead of having to log into a dedicated system to print the latest lab results or to view a recent x-ray, they can now use a tablet or a laptop to review the data in real time. They can also update the record at the bedside. The old adage that knowledge is power is very relevant here. If physicians can get test results in real time on their mobile devices, they can make faster, more informed decisions. Increased interaction with patients has been shown to increase the likelihood of revisits and ultimately patient loyalty. More time is spent at the bedside interacting with a patient if a clinician is not walking back and forth to a desktop machine, often in a different room.

Rather than trying to cover every scenario, one way to illustrate clinical workflow improvements is to tell two similar stories of a patient visiting an inpatient facility, where one of the facilities uses Wi-Fi heavily, while the other does not.

Mr. Xiou is a 39-year-old non-English-speaking Chinese gentleman who has recently experienced some irregular heart activity and has scheduled an appointment with his primary physician to look into the matter. His physician uses computers in the practice mostly for administrative tasks and has a stand-alone EMR system, but he does not leverage Wi-Fi. Mr. Xiou asks his spouse to drive him to the clinic. Upon his arrival, he is asked to fill out a paper form with his latest health information and the issue that he needs to see his physician about. He completes this in fairly quick order and submits it in Chinese. After realizing that he does not speak or write in English, the receptionist calls a language interpreter to come onsite to assist. The interpreter takes approximately 45 minutes to arrive. Upon his arrival, Mr. Xiou is ushered into the clinic by the receptionist and is led to an exam room. Meanwhile his spouse is by his side. His physician is able to see him after a half-hour wait. He looks over his history in the EMR system and decides that Mr. Xiou needs to have an EKG and blood labs. Mr. Xiou indicates that his wrist is also hurting him after falling on it a day earlier, so his physician orders an x-ray to be done as well to ensure that there are no fractured bones.

The physician insists that the tests need to be completed right away, and Mr. Xiou is shown how to reach the blood lab, the area where the EKG will be conducted, and the x-ray room.

Meanwhile one of Mr. Xiou's relatives is trying to reach his wife to check on him but she is unable to reach her due to poor cellular reception in the clinic.

After spending about 4 hours having the various tests completed, including waiting for the nurse to find an EKG device with wired EKG leads, having to go to a dedicated x-ray room, and having to wait for his turn in the lab, Mr. Xiou is back in the waiting room. His physician is waiting for the results from the various tests to be manually entered into the EMR system. Two hours later, he notifies Mr. Xiou that there is nothing alarming in the results, but that he needs Mr. Xiou to capture an EKG trace overnight, using a special recorder at his home.

Mr. Xiou agrees to this, and he leaves the hospital. When he arrives at home, he attaches the device and begins recording his EKG readings. His physician is unable to review the results until the next day when Mr. Xiou uploads them. After having more time to review his record in more depth, his physician notices that a combination of medications that he prescribed to Mr. Xiou a few days earlier may have been the cause of his irregular heartbeat, so he immediately calls Mr. Xiou and notifies him that he should not take the two medications at the same time.

This first scenario was a fairly positive visit to the clinic with the exception of the amount of time some of the testing required.

We consider a second scenario, in which Mr. Xiou has a physician who completely embraces mobility and has a wall-to-wall wireless network at his clinic which he uses wherever possible. Mr. Xiou asks his spouse to drive him to the clinic. Upon his arrival, the receptionist hands him a tablet PC to update his health information. The tablet is set up to accept input in various languages so Mr. Xiou is able to use his native tongue to enter his information.

After submitting his information, it is automatically sent to his physician's handheld device and updated in the EMR system. He is then ushered to the waiting room. His wife pulls out her smart phone and joins the wireless guest network after accepting the terms and conditions. She is now able to browse the Internet at will. Upon reviewing his information the physician realizes that Mr. Xiou only speaks Chinese and brings a workstation on wheels into the exam room. He launches a video conferencing session to a remote interpretation service, and is linked to an interpreter within a couple of minutes. In addition to the workstation on wheels, his physician has his tablet PC. He is able to review the EMR on his tablet and pull up Mr. Xiou's record. He looks over his history in the EMR system and decides that Mr. Xiou needs to have an EKG and blood labs. Mr. Xiou indicates that his wrist is also hurting him after falling on it a day earlier, so his physician orders an x-ray to be done as well. The physician notices an alert in the EMR indicating that Mr. Xiou is taking a combination of medications that can impact his heartbeat. He immediately notifies Mr. Xiou that he should not be mixing the two medications, but to be on the safe side he decides to move forward with the other tests.

His physician leaves the room and asks one of his assistants to conduct the test. The assistant locates the mobile x-ray machine, blood gas analyzer, and EKG device using the RTLS system interface on her tablet. She is able to locate the three devices in a matter of minutes. She then brings these to the patient's room, and within a matter of about an hour she is able to conduct all three tests. Since the devices are connected to the wireless network, and they have wireless sensors, a minimal amount of cables is required. The test results are then transmitted over the wireless network to an interface engine that integrates with the EMR. The physician is able to review the results as soon as they arrive. Within an hour, he notifies Mr. Xiou that there is no reason to worry, but that he would monitor his EKG overnight. His wife reaches out to their family and friends via e-mail over the wireless guest network to let them know that Mr. Xiou is OK. His physician provides Mr. Xiou with a device that utilizes wireless sensors to use at home. He also provides Mr. Xiou with a smart phone to capture the data and transmit it every 5 minutes to an application running on his smart phone. Mr. Xiou follows these instructions, and as his physician is having dinner he can see that the EKG results are normal and he is able to rest knowing that the likelihood that Mr. Xiou has a serious condition is minimal. The next day, he calls Mr. Xiou to let him know that his chart looks great.

The end result in these two scenarios is the same, but the second is streamlined and takes significantly less time than the first. It is less stressful for Mr. Xiou and his wife, and saves a tremendous amount of time and costs for the physician. These scenarios are based in a small clinic, but the time and cost savings can be more significant in larger hospitals, where it may take up to an hour to locate a mobile device without RTLS. These types of workflow improvements can sometimes make the difference between life and death in time-sensitive emergencies. They also contribute to improving disease management and preventable readmission costs. Costs associated with chronic diseases account for more than four-fifths of total healthcare expenditure, so any improvements in this area can have significant impact on the ever-increasing healthcare costs.

mHealth

When we start looking at rising trends within healthcare over the last 10 years, we cannot ignore mHealth. Analyst forecasts estimate the potential value of the mHealth market will be $4.6 billion by 2014. Almost one in four adults in the United States reported using their smart phones to access health information in 2011. This includes looking up health information sites like WebMD and looking at their health records. These numbers are impressive considering that the first iPhone was released in June of 2007, and the first Android device was released in October of 2008.

One aspect that has been eluding health care professionals is a formal definition of mHealth. For the purposes of this book, we will use the definition created by the National Institutes of Health: "the delivery of healthcare services via mobile communications devices." We may also reference a more in-depth definition by the mHealth Alliance, which defines mHealth as medical and public health practice supported by mobile devices.

Global aging, rising healthcare costs, and a shortage of qualified health professionals are three of the key drivers for healthcare professionals to look at creative, new models of care like mHealth. mHealth encompasses a wide range of use cases that can be delineated around the physical boundaries of a traditional hospital. Mobility within hospital or clinic walls is very different from mobility outside the walls of the hospital. Some of the use cases for mobility and mHealth within a healthcare facility have been covered in the section on workflow improvement. Once we start scrutinizing use cases outside the traditional four walls of the hospital, we start to look at technologies beyond Wi-Fi.

Global cellular subscriptions have grown from 962 million subscribers to 5.9 billion subscribers, and the trend is still increasing.

With 90 percent of the world's population covered by some form of cellular signal, the use cases quickly increase.[7] These can range from patients using the cellular network to send their physician live physiological measurements, to physicians remotely diagnosing patients in rural areas that are greater than walking distance from a hospital. Developed countries take advantage of this type of technologies in different ways than developing countries. Text messaging is the primary form of communication between physicians and their pregnant patients in Nairobi, whereas in the United States this can include video or voice communications.

The number of health-related mobile applications has increased dramatically over the last 5 years. These target wellness, EMR access, and patients' curiosity about medical symptoms.

Endnotes

1. "Hedy Lamarr: Inventor of More than the 1st Theatrical-Film Orgasm.". *LA Times*. November 28, 2010.
2. N. Abramson (1970). "The ALOHA System: Another Alternative for Computer Communications." *Proc. 1970 Fall Joint Computer Conference* (AFIPS Press). November 17–19, Houston, TX.
3. Garritty, C., and K. El Emam (2006). "Who's Using PDAs? Estimates of PDA Use by Health Care Providers: A Systematic Review of Surveys." *J Med Internet Res* 8(2):e7. http://www.jmir.org/2006/2/e7/
4. http://www.wi-fi.org/files/WFA_Timeline_Updated_PDF.pdf (accessed June 10, 2013).
5. http://www.fcc.gov/encyclopedia/wireless-medical-telemetry-service-wmts (accessed June 10, 2013).
6. Tara Moore, "Spectrum Squeeze: The Battle for Bandwidth," *Fortune Magazine*, July 27, 2011.
7. Chris Sweeney, "How Text Messages Can Change Global Healthcare," *Popular Mechanics*, October 24, 2011.

2

WIRELESS ARCHITECTURE CONSIDERATIONS

About Wi-Fi Networks

Wi-Fi networks are a marriage between client devices and access points, which are referred to as stations (STAs) and APs, respectively. These entities have a few things in common yet maintain some clear differences. The APs communicate regularly with everyone around him or her, making sure that the STAs have the correct time and notifying the STAs when it is time to wake and do some work. STAs, on the other hand, are either receiving and sending packets or notifying the APs that they are going to sleep to conserve some energy. This analogy is meant to entertain but at the same time drive home the principle that a successful network design balances the needs of the AP and the STA client devices. Let's start with the AP. The AP is what creates a BSS or Basic Service Set. This is kind of like a date. It is just a concept not an actual measurable entity. If the BSS is extended over several APs, then we can call it an ESS or Extended Service Set. This is a little more like a marriage than a date. An ESS allows the STA to maintain a relationship with the network regardless of which AP it is communicating with. So, the BSS and ESS are a concept that is central to wireless local area network (WLAN) principles. Every BSS will have a BSSID, which is a media access control (MAC) address, of the APs radio and may look something like this: 00-c0-ff-ee-12-34.

For the past 10 years or more WLANs have undergone tremendous change. As with any new technology researchers and vendors are always inventing ways to improve the value of the core technology. For wireless LANs and really any networking technology these changes fall in one of three "planes." Networking infrastructure is commonly described as having a management, control, and data plane. Each plane serves a different function. The management plane typically is concerned with configuring and monitoring the wireless infrastructure. The control plane is primarily concerned with how to handle the data plane. The data plane is just what the name implies, the bits and bytes of an email, web page, or other application that rides over a network. Ideally all three of these planes would be completely implemented in the access point. Earlier APs did, in fact, implement all three planes in the access points. Unfortunately, every device needed to be configured and managed individually. This is why these APs are called autonomous. Some autonomous access points could be managed centrally via protocols such as SNMP, Telnet, and SSH.

The MAC Layer

The 802.11 standard uses two layers out of the Open Systems Interconnection (OSI) model, the physical and data links. The data link layer has a sublayer called the MAC or media access control layer. This is where the majority of the 802.11 protocol lives. There is also a physical layer but this mostly lives in the silicon of the radio chips that make the wireless transmissions possible. This is a very in-depth discussion that has no place in this chapter. The remainder of this chapter discusses the MAC layer and its interactions with higher layers like the network layer. As its name implies the MAC layer provides access to the media. In the case of 802.11 networks the media is the radio frequency spectrum that the AP and client will use to communicate with each other. The MAC layer functionality can reside on the AP, wireless controller, or be split between both devices. Where this functionality resides may impact your network design.

MAC layer variations include (Figure 2.1):

1. Local MAC—all MAC functions on the AP
2. Remote MAC—all MAC functions on the controller
3. Split MAC—Real time (RT) on AP and non-RT on the controller

Over the last 10 years the control of the MAC layer has shifted from the AP to a controller and now is moving back to the AP. The location of the MAC layer functions depends on what problems need

802.11 Functions	Local MAC	Split MAC	Remote MAC
Controller Functions (Control)	Controller	Controller	Controller
802.11 MAC Non-Realtime		Controller	Controller
802.11 MAC Realtime	AP	AP	Controller
802.11 PHY	AP	AP	P

Figure 2.1 MAC scenarios.

to be solved. Each of the above MAC scenarios could then be mapped to the following vendor-specific solutions.

Vendor-Specific Solutions

1. SWAN—Structured Wireless Aware Network (Cisco)
2. LWAPP—Light Weight Access Point Protocol (Cisco)
3. SLAPP—Secure Light Weight Access Point Protocol (Aruba and Trapeze)
4. PAPI—Process API (Aruba)
5. CAPWAP—Control and Provisioning of Wireless Access Points

Most vendors tunnel the wireless client traffic through what is called an IP in IP tunnel. IP stands for Internet Protocol. You may or may not be familiar with the technicalities of protocol but we can guarantee that you use it every day. IP in IP tunneling takes an entire data packet and stuffs it into another one. In this way data can traverse routers unchanged. There are some reasons why this was and is desirable. Many early attempts were made to standardize the tunneling method. In theory, one vendor's controller could interoperate with another vendor's AP. This way APs from different vendors could be deployed seamlessly on the same network. Some examples of this were LWAPP, CAPWAP, and SLAPP, listed above. Unfortunately, none of the proposed standards were robust enough on their own, and vendor-specific implementations emerged. This has led to more proprietary tunneling methods. We expect to see only control traffic in proprietary tunneling from now on. Furthermore, the control plane will likely move from the controller to the AP. There is a general trend to move to a distributed architecture, which places the data, control, and management planes directly on the AP and eliminates the need for a controller. The primary driver for this move is the increased data rates from 802.11n. The advent of 802.11n created a bottleneck at the controller. Twenty 450-Mbps 802.11n APs could easily overrun a wireless controller. Future technologies with even higher speeds could push this issue further. Let's start at the earliest widely deployed architecture, the autonomous AP or fat AP.

Autonomous Architecture   In the beginning all APs were simply called APs. Today we call them autonomous or fat., but only if your personal ethos permits such cruelty. However, once vendors released new controller architectures they needed something to call the old devices that they were replacing. These devices typically have the fewest number of features out of all the architectures. Autonomous APs are simply IEEE 802.11 to IEEE 802.3 bridges or Wi-Fi to Ethernet bridges. A bridge simply converts one protocol to another. Some of my earliest experiences were with IEEE 802.11 to IEEE 802.5 bridges.

IEEE 802.5 is the token ring protocol, which will not likely ever be seen again outside of a lab. They simply took an Ethernet or possibly token ring frame, removed the Ethernet header part and replaced that header with an IEEE 802.11 header and added an FCS or Frame Check Sequence at the tail of it. The WLAN's role at this point was very minimal. The management, control, and data planes resided on each individual AP. Each AP is configured individually and forwards data to the switch it is connected to. These devices were not capable of IP routing or IP mobility.

Once higher Physical Layer (PHY) (radio frequency) rates began to emerge, such as 802.11b, 802.11a, and 802.11g, it was possible to connect more devices to one AP. An autonomous AP would typically have one or more SSIDs dedicated to some sort of corporate access and possibly one for guest access that was unencrypted. Each of these SSIDs would be bridged to a particular Virtual Local Area Network (VLAN) to segment each access type. Typically a VLAN would have a subnet associated with it. Subnets allow for network nodes to talk to each other without requiring a router modifying the frame. Subnets have something called a subnet mask that defines how many nodes will be on the subnet. A mask of 255.255.255.0 or something like that which would allow for only 256 IP addresses minus the network, broadcast address, gateway, and any other management addresses would typically leave at least 250 addresses left for wireless clients; 256 addresses per subnet has been a best practice to limit the quantity of broadcast traffic sent to every device on the LAN.

This isn't as much of a problem for the switches and APs as it is for the clients. Every broadcast frame that is transmitted over a LAN switch has a destination address that everyone must process. While enterprise-switching equipment has more than enough horsepower to forward every broadcast frame it receives, a client Ethernet NIC will not always have the same capability. Large amounts of broadcast frames on a LAN can begin to impact application performance. This is primarily a problem that affects wired devices. However, broadcast traffic does in fact affect wireless devices but in a different way. Every broadcast frame that is transmitted on a subnet is sent to every AP on it. Then every AP on the same subnet forwards this to each radio interface associated with the subnet at the lowest data rate. This is a terribly inefficient use of wireless spectrum and why subnet segregation may be a good idea. As the wireless revolution began it didn't take too long for all the IP addresses in the subnet to become completely filled up and helpdesks were inundated with calls complaining that "my computer cannot get to myspace.com." It is possible to make the IP scopes larger but we just discussed why you might want to avoid this solution. You could put the APs in different VLANs with different IP scopes. But people would complain that their applications would fail when they changed subnets and if your wireless client needs a static IP address, then you are just out of luck. Some implementations could add more SSIDs to the AP. This only created new problems as they solved the old ones. If only there was a magic box that aggregated all the WLAN traffic and fixed all these issues. Somebody queue the music as the WLAN controller rides in on its white horse.

Controller-Based Architectures The WLAN controller made it possible to have a full featured yet still scalable WLAN. Controllers moved the control, management, and data planes to a centralized location. This allowed the administrator to configure and manage just a few controllers even if they had thousands of APs. One version of Cisco's LWAPP architecture encapsulates an 802.11 wireless frame in an Ethernet frame and sends this to and from the WLAN controller.

This is called local architecture as the MAC functions are on the controller. The AP was simply a radio antenna. All of the intelligence was centrally located on the controller.

Functions like 802.11 Beacons were handled by the AP and did not require controller intervention. However, functions like Authentication and Association were handled by the controller. Not all vendors took the same approach. Many other vendors processed the WLAN header at the AP and bridged the payload into an encapsulated tunnel. Tunneling is a common practice in controller based WLAN equipment.

Controller to AP Tunneling allows the WLAN client traffic to be segmented from any other traffic on the wired network. This solves many problems like having to span VLAN's across edge network closets. Doing so would open the possibility of having very large broadcast domains. A large amount of broadcast traffic is known to degrade the performance on wired LANs and WLANs. When the client traffic is tunneled to a controller, the effects of a large broadcast domain can be minimized. The WLAN controller owes nearly all of its success to the rapid adoption of Wi-Fi.

Where do controllers come from? Controllers were forged out of the need to solve a problem with a limited toolset. In the case of the WLAN, almost every problem that a controller is attempting to solve is rooted in having higher and higher densities of wireless clients. This is saying that there was nothing wrong with autonomous APs until many more of them were put together, and a hoard of clients began using them. This is all about scalability. As more clients are added to a network, the complexities increase. This requires the AP developers to consider these new problems and try to fix them with new software. This is only where the problem begins, and it is a slippery slope. As more functions are crammed into the AP's operating system it may begin to affect its performance as well. So there may be a balancing act even as architectures migrate towards the distributed model. The software developers have a meeting with the hardware developers. The conversation goes something like this.

Software guy says: "We need a faster processor."

Hardware guy says: "OK, how much faster?"

Software guy says, "Hmm, maybe twice as fast? But wait, it needs a five-year lifecycle, and how about quadruple the speed?"

Hardware guy says, "The only processor I have that supports real-time embedded OSs uses more power than the 15.4 watts that power over Ethernet provides."

The hardware and software guys agree that according to Moore's law it will be 36 months until they have the processing power necessary. In the meantime, they will offload the extra processing to a central controller until embedded technology catches up with the market's needs. Not only do AP processors need to be fast they need to be efficient as they get their power from an extremely finite resource. IEEE 802.3af or power over Ethernet specifies a maximum of 15.4 watts available for consumption. IEEE 802.3at effectively doubles the power. A modern computer consumes about six times that amount just for the CPU. Designing a robust and fully functional AP is a bit like packing for a trip to Mars; only pack what you absolutely need. Don't pack a server cluster in your spaceship if mission control can provide you the same information in an acceptable amount of time. 802.11a and 802.11g accelerated the adoption of controllers since the AP now had to process nearly ten times the traffic as it did with just one 802.11b radio. In order to meet the processing requirements of the AP many functions were offloaded to the controller. Once again 802.11n added nearly a tenfold increase in throughput processing.

Controllers take care of many functions, such as IP roaming, authentication, log aggregation, encryption key distribution, mitigating broadcast traffic and upgrading access points. Almost all controllers tunnel the client traffic back to the controller for de-encapsulation. This adds overhead and complexity to the solution but improves scalability. In the healthcare industry, wireless networks have made a traditionally technophobic workforce overnight Wi-Fi mavens. Without the role of controllers in the healthcare industry adoption of the sheer volume of devices would have been much more difficult and problematic. About the time 802.11n found its way into the enterprise, it started becoming inconvenient to have a controller in your data path. With APs having up to 900 Mbps radio throughput, it is obvious that even a controller with 10 Gigabit interfaces would at some point become the bottleneck. With this, virtually every vendor has begun to solve a new scalability problem: how to handle all of this traffic? Fortunately, embedded CPUs have continually increased speed and new ideas about the data plane have emerged. Wi-Fi vendors are focused on how to move their data plane out of the controller. Many are also looking at and even implementing ways to move their control and management plane out of the controller. This could be regrettable with new technologies like 802.11ac and 802.11ad emerging. I would not be surprised if the WLAN controller reinvents itself as something like a private cloud computing system as APs once again become very resource constrained under the new throughput load. Every time a radio chip increases its throughput, the AP's processor must also do the same. Imagine that you attempted to put a 300-Mbps Wi-Fi card in your 386SX computer from two decades ago. It would not matter even if the turbo button is on, there is no way that a 16 MHz processor will handle 300 Mbps of Wi-Fi traffic. This is a reason that the modular APs radio did not last long. APs just don't have the same design criteria as a modular router.

The top problems controllers attempt to solve are

1. Subnet roaming
2. Performing the RADIUS Authenticator role
3. WPA2 key distribution and caching between APs
4. Proxying ARP
5. Provisioning Access Control Lists (ACLs) on every AP
6. Converting Multicast to Unicast
7. Layer 4 traffic inspection
8. Guest captive portals

Distributed Architecture   Distributed architecture is the new kid on the block. 802.11n challenged the idea of being able to centralize the data plane. I do not believe controllers were ever intended to be a permanent solution. It was either impossible or cost prohibitive to implement the software that the designers wanted. That said, it is very possible to implement more functionality in modern AP's and it is more cost effective. New APs are being developed that solve most of the problems that designers intended the controller to solve. These new APs provide the hope of higher forwarding throughput and better overall performance for the client devices. Distributed APs have some new problems to solve. First, how do you manage all these devices? With either a clever software client and AP configuration distribution software or Cloud-based management plane. What is the line in the definition between a controller-based and a distributed architecture? In the purest sense of the definition all data, management and control functionality and software would reside on the AP. Autonomous APs sound like they meet this definition; however, I would add that the distributed APs should have a single point of management. Since autonomous APs relied on individual management or a network management suite, this would rule them out as being distributed. This architecture certainly poses some challenges.

Single-Channel Architecture (SCA) versus Multichannel Architecture (MCA) There have been some vendors that use single-channel architecture (SCA). SCA just as it says only leverages a single Wi-Fi channel for the entire network. This is accomplished by making every AP look like the same AP to the client. The advantage of this solution is to virtually eliminate roam times between APs. In addition, there is no need to provide any channel planning or leverage automatic channel selection. The downside is that you only have the bandwidth of one channel available. One key driver of this solution is VoWiFi. Voice traffic requires roam times below 50 ms and generally is a low-bandwidth application. While this strategy has been implemented in a number of deployments only two vendors to date have implemented SCA in their products. The remainder of this book will focus on multiple-channel architecture or MCA.

Building for Applications and Performance Designing a wireless network is much like writing a book; you begin with identifying the audience. Mobile devices are the audience of the WLAN. As such, it is incredibly important to take into consideration all of their nuances and applications. Devices range from smart phones to intravenous pumps, each of which creates unique challenges for a wireless engineer. One such problem is static IP addressing. This is a perplexing enigma. Why would a device that moves around have a fixed IP address? Whatever the reason is, many medical devices will require the use of static IPs and there are solutions for this.

First let's talk about why this might be a problem. In smaller networks, this is typically not a problem. Regardless of the network size, as the number of clients increases, so does the need to segment their broadcast and multicast traffic. So, regardless of the number of access points, large numbers of clients will stress a single Extended Service Set (ESS).

Unicast, Multicast, and Broadcast   IPv4 was originally designed for wired networks. Many functions of the IP protocol are not ideally suited to the 802.11 protocol. The 802.11 designers did their best to work with IPv4, as it was the de facto standard. IPv4 has three types of frames, unicast, multicast, and broadcast. Unicast frames are sent from one address to another address. Over the wireless medium, the unicast frames are sent at the highest PHY rate that the client selects based on a number of criteria. By contrast multicast and broadcast traffic are flooded out to every device on the same subnet. On a WLAN, these packets are flooded out of every AP in the same IP broadcast domain. Unlike the unicast frame, these are transmitted at the lowest supported PHY rate. On an 802.11g network, this is the difference between as much as 54 Mbps and 1 Mbps. The broadcast 802.11 frames requires as much as 54 times as much transmission time as the unicast frame. If enough clients are producing these emissions, even infrequently, this could create severe performance problems on the WLAN. Many problems that can be found on wireless networks are an army of mice rather than a fire-breathing dragon. Figures 2.2 and 2.3 show two different packet captures of client devices sitting idle. At first glance this would seem inert. However, if 500 clients multiply this traffic, it is suddenly alarming. So, it is imperative to minimize this traffic whenever possible.

The two figures show the same device under different circumstances. The x axis is time in seconds, while the y axis represents the number of packets.It may be surprising that the graph in Figure 2.2 is of a phone associated with a Wi-Fi network with the screen turned

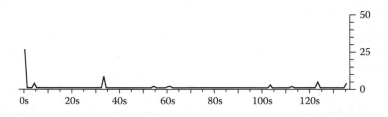

Figure 2.2   Idle associated iPhone traffic graph.

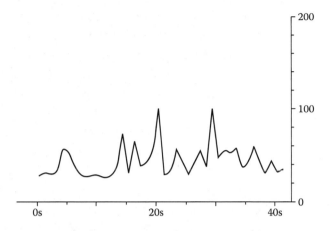

Figure 2.3   Idle unassociated iPhone traffic graph.

off and the phone in an idle state. The phone in this state seems to peek at about 5 packets per second. Even with 300 phones on the same network this amounts to 1500 packets per second. Any modern wireless LAN should barely notice this amount of traffic. Nearly all of this traffic is being transmitted at higher data rates, as this traffic is data wireless LAN traffic. Data traffic is sent out at what the client believes to be the highest data rate at which the AP can reliably receive the traffic. Figure 2.3 shows significantly more traffic. It seems to be averaging about 50 packets per second and peaking at 100. What is surprising is that this is the same phone with no Wi-Fi connections available. Even at 10 to 20 times the number of packets per second, this may not seem too bad. However, there is a hidden story. The packets in the first graph (Figure 2.2) were transmitting at data rates as high as 65 Mbps, while the packets in the second graph (Figure 2.3) are being transmitted at a paltry 1 Mbps. In Wi-Fi networks, the number and size of packets is significantly less important than the data rates at which they are transmitted. Data rates correspond to the amount of "airtime" that a client uses. This is sort of like how many minutes you use on your cell phone. The number of calls is less important than the total minutes used. Just check your cell phone bill and you will be more concerned about the number of minutes used rather than the number of calls made. Cell phone companies charge by the minute because the wireless spectrum they use is a finite resource. In much the same way Wi-Fi networks have a finite amount of spectrum. If we add up this equation with this new understanding we paint a completely different picture about this disruption that this one unassociated device is creating. If this device is transmitting ten times as many packets at 1/65th the data rate, we arrive at a factor of 650. The unassociated client in the second graph is utilizing up to 650 times as much airtime or bandwidth as the associated client in the first graph. In other words, you could have 650 idling clients connect to your wireless AP with roughly the same spectrum bandwidth consumption as one unassociated client. So remember, the number of calls is less important than the call length.

Now back to the issue of static IP addresses. Unfortunately, the solution to this problem varies with each WLAN vendor. However, most vendors will have a solution that leverages a single VLAN that can span the entire wireless network. In this solution a single VLAN is employed by one of several methods. The first method is to use a Role-Based Access Control or RBAC to assign the VLAN to the individual client. This VLAN can be applied by the AP at the edge of the network or by a controller at or near the core of the network. Again this will vary with each vendor. In this scenario, the WLAN has a single SSID to manage for multiple device types. The broadcast domain can be minimized through the use of smaller subnets. The downside is that a RADIUS server must be managed to provide the RBAC rule set that assigns the VLAN.

Another option is to create a dedicated SSID for each subnet. While this solves one problem it creates another. Suppose that each subnet is set to a 24-bit subnet mask (255.255.255.0), which is very common. This will limit the number of devices that can use this subnet, to 254 to be exact. However, as the subnet size increases so does the amount of broadcast and multicast Ethernet frames. As I have mentioned, this becomes increasingly dangerous as all these Ethernet frames are flooded out every AP radio that is connected to the VLAN. While there is no standard by which to gauge the best practice subnet sizing, it is recommended to keep your Wi-Fi broadcast traffic under one percent. There are two reasons for following this guideline. First the low data rate broadcast traffic will greatly reduce the aggregate throughput of the WLAN. Second, high amounts of broadcast traffic on a WLAN will make troubleshooting an application performance problem very difficult.

There are some options available to combat Ethernet broadcast and multicast traffic. One specific broadcast Ethernet frame called an ARP frame or Address Resolution Protocol frame can create a significant amount of Wi-Fi broadcast frames. Fortunately many wired and wireless companies implement something called proxy ARP. RFC (Request for Comment) 1027 or proxy ARP does just what its name suggests. It will allow another device on the same network as an end client to "ARP" for another. The purpose of ARP is to update the ARP tables on all the devices on a LAN segment. An ARP table is a mapping of MAC addresses to IP addresses. Switches and end clients alike maintain some level ARP table. You can observe this by typing the following in the command prompt of a Windows PC:

```
C:\arp.exe -a
```

This yields something like this:

```
Interface: 192.168.2.215— - 0x14
  Internet Address     Physical Address     Type
  192.168.2.1          00-24-01-39-9c-7d    dynamic
  192.168.2.2          b8-27-eb-0f-ad-b1    dynamic
  192.168.2.192        a4-d1-d2-23-11-2a    dynamic
  192.168.2.194        d8-d1-cb-74-6a-db    dynamic
  192.168.2.205        7c-e9-d3-b2-6a-34    dynamic
  192.168.2.208        d0-23-db-94-c7-02    dynamic
  192.168.2.209        00-26-ab-49-72-f7    dynamic
  192.168.2.255        ff-ff-ff-ff-ff-ff    static
  224.0.0.2            01-00-5e-00-00-02    static
  224.0.0.22           01-00-5e-00-00-16    static
  224.0.0.251          01-00-5e-00-00-fb    static
  224.0.0.252          01-00-5e-00-00-fc    static
  239.255.255.250      01-00-5e-7f-ff-fa    static
  239.255.255.253      01-00-5e-7f-ff-fd    static
  255.255.255.255      ff-ff-ff-ff-ff-ff    static
```

This output shows the mappings of IP addresses in the leftmost column to the MAC address in the middle column. And the right column indicates the type of ARP entry. In order to maintain this mapping of IP addresses and MAC addresses each device will send out not only ARP frames for itself but also solicit responses from other devices. This can create a tremendous amount of broadcast traffic. In fact in the older 10/100 fast Ethernet networks as few as 16 broadcast frames per second would begin to degrade application performance simply because every client device must process the broadcast request. This requires the client to stop whatever it is doing to listen to the broadcast frame, possibly look up information in RAM or storage, and possibly send a response. As you can see broadcast/multicast traffic can unnecessarily tie up wireless bandwidth and client device CPU cycles. Proxy ARP prevents end clients from being interrupted. The way that this technology works is typically by capturing the Dynamic Host Configuration Protocol (DHCP) transactions, specifically the DHCP-ACCEPT packet. Any device on the network that performs a DHCP relay function would be able to see what IP every MAC address has for this local network. APs, controllers, and switches typically will perform the DHCP relay function. This allows one of these devices that are in line with the end client to not forward the ARP request to the client and instead answer back to the ARP frame originator. This simple function can save your clients a significant amount of work. This translates to fewer CPU cycles, less bandwidth utilization, and as a result a higher battery life in the case of mobile devices.

The second measure that can be implemented is converting broadcast traffic to unicast traffic. Not every vendor implements this so you may have to consult your vendor documentation. The primary effect this solution has is to reduce the amount of time that the wireless medium is busy. This may seem counterintuitive that it could take less time to transmit say the same packet 20 times to 20 different clients than it would take to send one frame to 20 clients. Let me explain a little further. Every client on an ESS or BSS is able to decode a broadcast frame and must at least read the header to verify whether it is intended for it. Broadcast frames on an encrypted network use a common key among all the clients called a Group Temporal Key (GTK). This is what allows all the clients in the BSS to decode a broadcast frame if it is encrypted. However, broadcast WLAN frames are not acknowledged. This is known as the two-way handshake. Unicast frames are usually followed by an ACK or Acknowledgment frame to indicate that the data was successfully received. It would take even more "airtime" to have every client connected to the BSS ACK every broadcast. So, the 802.11 standards do not specify that broadcast frames be acknowledged. Now, if you were in a room full of people and you wanted to ensure that everyone in the room was able to hear everything you were saying without repeating yourself you would do two things: speak loudly and slowly. This is exactly what happens with broadcast frames. Broadcast frames are transmitted at the lowest basic rate. The basic rates in a typical WLAN would be 1, 2, 5.5, or 11 Mbps for 802.11b and 6, 12, 18, or 24 for 802.11a or 802.11g. This is a magnitude slower than 450 or even 300 Mbps maximum unicast frame rates. One broadcast frame at 1 Mbps would take just as much "airtime" as 300 different frames at 300 Mbps. Even one broadcast frame at 24 Mbps, the maximum basic rate, would occupy the same amount of airtime as 12 unicast frames transmitted at 300 Mbps.

One of the most common sources of multicast traffic in a hospital is Vo-Fi or Voice over Wi-Fi. Phones will use multicast frames for functions like registration and Push-to-Talk (PTT). There are some unique challenges with providing this functionality. PTT functionality allows workers to use their phones in an intercom fashion to communicate with a group of people. As hospitals have become more cognizant of their patients' mental health and comfort, overhead paging has slowly begun to disappear from the hallways and patient rooms. PTT emerged as a compromise between the overhead paging and none at all. The problem with multicast groups is that they are only transmitted on the local subnet of the transmitting device. There are ways to route multicast traffic. However, this introduces new complexities to the network architecture. The simplest solution to this design requirement is to simply place all the clients that are required to use multicast on the same subnet. If there are a significant number of devices this may require increasing the subnet size. Large subnets are not a problem as long as the risks associated with creating these are understood and measures are taken to minimize the broadcast traffic.

Hopefully by now it is becoming very clear that broadcast traffic typically does more damage than good to your WLAN and should be avoided unless it is necessary to meet a business requirement such as PTT. Much like the wired networks in past years, WLAN administrators must be diligent to find ways to minimize and eliminate broadcast 802.11 frames. Leveraging technologies such as proxy ARP and multicast to unicast translation is a good starting point to improve the efficiency and usability of your WLAN.

Medical Devices

Medical devices often pose unique challenges when designing a WLAN. Medical devices are nothing new to a hospital. In fact many leveraged early technologies like serial communications and modems to transmit medical data back to a main frame. Unfortunately, some medical device vendors have not moved much beyond this communication paradigm. I have seen my share of devices that use serial to Wi-Fi bridges. There are some fundamental design flaws with this approach. First, one would beg to ask why use serial? The inconvenient truth about this is that it was cheaper to reuse an old product and bolt on the wireless capabilities than to redesign the entire product to have Wi-Fi built in. Also, serial may actually have enough throughput to transmit the necessary data. The good news is that even high speed serial is relatively slow compared with modern 802.11n networks. In a nutshell, this traffic does not necessarily pose a capacity concern like streaming HD video. These types of solution are the most common ones that require static IP addresses, and lower encryption standards such as Wired Equivalent Privacy (WEP), Temporal Key Integrity Protocol (TKIP) encryption. They are also likely to use only 802.11b/g, which utilizes the heavily saturated 2.4-GHz wireless spectrum. Again, a serial interface will not have very high data rate requirements.

One important thing to note is that most test equipment will not be mobile, but rather nomadic. The difference is in semantics. A mobile client will require data usage during transit. Examples of this are Wi-Fi phones, real-time location tracking, and supply chain delivery robots. Nomadic devices are moved throughout a facility but are only used in one place at a time. Equipment such as intravenous pumps, x-ray carts, ultrasound machines, pulse oximeters, and vital sign monitors. While these devices may travel throughout a hospital, they are seldom transmitting on the wireless LAN during their travels. One important thing to keep in mind when deploying wireless medical equipment is to ensure that manual backups such as audible alarms are present in case of a network outage. Most devices will store information locally and simply upload the patients' data when a connection is available. This may be on a defined interval or a constant stream of data. That makes most medical devices just like a data application on any typical end user device. There are robust Transmission

Control Protocol (TCP) timeouts and retransmissions to take care of any intermittent connectivity issues. Most wireless medical devices do not require real-time communications, which requires timely delivery of data.

Medical Imaging

Medical imaging is quite possibly one of the most important tools that doctors have at their disposal. It affords the ability to look inside a person without making a single incision. This technology is the past, present, and future of medicine. I would wager that we will only see more imaging technology in hospitals and clinics. However, one of the key challenges sometimes is simply moving a patient to get an x-ray or ultrasound. Mobile units are the solution to minimize the risks of moving a low mobility patient. What is even better is when the caregiver can upload the image to a radiologist or other specialist and receive timely feedback at the patient bedside. Wi-Fi is the tool of choice for this solution. However, here is the challenge. Medical images are often quite large, and they are only getting larger with three-dimensional capabilities and video recordings. There are some key technologies to leverage on the WLAN and clients to improve the transfer speed of medical images. The following are Wi-Fi technologies that will improve the end user experience.

- A-MPDU
- A-MSDU
- Block Ack
- Contention Free Burst (CFB)

A-MPDU and A-MSDU are acronyms for Aggregated MAC Protocol Data Unit and Aggregated MAC Service Data Unit. What is important about these features has to do with the aggregated part. A-MPDUs and A-MSDUs allow the client devices and APs to take multiple packets and send them all at once. You may be asking yourself at this point "Wi-Fi doesn't work that way by default?" The answer is obviously no. Imagine your WLAN is like a traffic intersection with a four-way stop. Every car that comes to that intersection must stop before proceeding through the intersection. This is the default behavior of a WLAN. Every device must wait its turn and send one packet at a time. A-MPDU and A-MSDU allow multiple packets to be lumped together as one large packet. This improves the overall efficiency of the WLAN and allows medical images to transfer faster. This would be like loading ten cars onto a car hauler. When the car hauler comes to the intersection, only the hauler must wait for its turn. All the cars it is carrying will be allowed to pass without waiting. Block Acks or Block Acknowledgments were introduced with the 802.11e standard and are another great improvement to the efficiency of the 802.11 standard because it improves what is called the two-way handshake of Wi-Fi data transmission. After every packet that is transmitted the AP or client needs to verify the transmission made it successfully. Block-Acks allow the AP or client to avoid performing the two-way handshake every time. Block-Ack allows several packets to be acknowledged with only one ACK frame.

Transmission sequence without block acknowledgment

Transmit data → ACK → Transmit data → ACK

Transmission sequence with block acknowledgments

Transmit data → Transmit Data → Transmit Data → Block ACK

Finally, a CFB or contention-free burst allows a client or AP to transmit for a defined duration without contention during a TXOP or transmit opportunity. Going back to the analogy of the traffic intersection, CFBs would be like a four-way intersection with a two-way stop. Wi-Fi Multimedia (WMM) defines four access categories (ACs): voice, video, best effort, and background. Only the voice and video ACs are allowed to participate in the CFBs. This is valuable for video streams but not as much for voice. Video packets usually are large and numerous. A CFB will definitely increase the efficiency of the WLAN. However, it may not actually improve the visual experience. Hollywood movies are typically recorded at 24 frames per second. Many have accepted 30 frames per second as the rate at which the human brain does not see the individual frames. Either way it means that a frame change occurs every 33 milliseconds (ms) to 41 ms. With a 1500-byte maximum Ethernet payload and an average data rate of 1 Mbps for a 720 p HD video stream, that would correlate to about 683 packets per second. This would indicate a 1500-byte packet every

1.5 ms. If you have 33 to 41 ms to transmit one frame change, CFBs will definitely improve your odds. Voice traffic has a similar math equation to solve. CFBs will aid in the timely delivery of voice traffic as well. Voice over Wi-Fi is discussed in detail in Chapter 7.

Wireless on Wheels

Wireless on wheels or WOWs are nothing more than a laptop or desktop computer mounted to a cart. WOWs have historically leveraged fairly lightweight applications that are tolerant to delay. Therefore, application delivery on a laptop is not as difficult as other real-time application. A laptop has significantly more processing power than a mobile device such as a tablet or smart phone. There are some considerations to keep in mind when designing Wi-Fi networks for WOW carts, which have more to do with how they are used than the devices themselves.

First, they can be used rather sporadically for things like physician rounds or during system updates. Doctors often do rounds in the morning after their patients have had all their vitals taken and some breakfast. In addition teaching hospitals will have a small army of residents following the lead physicians around. It may not be uncommon for everyone in the entourage to carry around a WOW cart. These high densities can pose problems for other high availability applications like voice. Sure QoS measures can improve performance of a specific application. However, many engineers are not aware that Wi-Fi, QoS, WMM, and IEEE 802.11e only improve the statistical odds that your critical application's traffic gets priority. Modern laptops will likely have high-performance Wi-Fi chips. One solution to help with this problem is to ensure that minimum basic rates are set as high as possible, and that the highest data rates are available by enabling 40-MHz channels on the 802.11n 5-GHz radios. Increasing the minimum basic rates can reduce the airtime requirement for management and control traffic to support high densities of clients by 95 percent. It is necessary to take this into consideration during the planning phase and site survey to be successful with this strategy. If your network is not designed from the beginning to support high-density clients, taking these measures will only create problems. Surveying for high-density clients is covered in detail in Chapter 3.

Second, when WOW carts are found in storage or charging areas, they may be associated with the same AP. For example, ten WOW carts connected to the same AP receiving a 500-MB patch will go unnoticed on the carts themselves, but the patient monitor or Wi-Fi VoIP phone may have degraded service due to the airspace contention. In addition, WOW carts may wake up and connect all at once to the WLAN, possibly depleting the available DHCP leases available. Creating dedicated SSIDs or VLANs for WOW carts will help mitigate the problems described here. Some key facts to keep in mind when dealing with WOWs are:

- Can be a laptop or all-in-one PC
- Can be used for EMR applications
- Modern laptops have 3x3:2 and 3x3:3 MIMO Wi-Fi chips
- Usually support OKC and Pre-Authentication for fast roaming

Two new types of devices have emerged after 15 years of solid PC growth, the tablet and smart phone. These devices are significantly different from their larger, bulkier counterparts. Mobile tablets and smart phones will continue to grow in popularity and consume market share from the aging laptop format; however, this transition does not come without some trade-offs. The biggest and most obvious trade-off is real estate. A laptop computer over the years became increasingly comparable to the even older desktop computer in computational power and storage. The two major constraints on a mobile device are real estate and power. As such, the Wi-Fi radios that are present on these devices are very different from the ones on their laptop cousins.

A typical modern enterprise-grade laptop computer is capable of 300 to 450 Mbps. This is a stark contrast to the smart phones and tablets that usually support only up to 65 Mbps. These lower data rates are the result of a compromise on real estate and, mostly, power. The 802.11n standard specifies data rates up to 600 Mbps. This is accomplished through the use of multiple radios and antennas. Mobile device designers must balance the performance and power requirements of the device. As such Wi-Fi radios on these devices are minimally featured. Furthermore, most mobile devices have limited storage capabilities and therefore may not need the ability to transfer

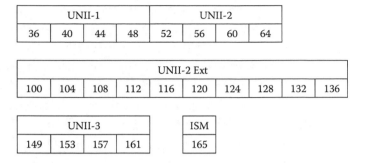

UNII-1				UNII-2			
36	40	44	48	52	56	60	64

UNII-2 Ext									
100	104	108	112	116	120	124	128	132	136

UNII-3				ISM
149	153	157	161	165

Figure 2.4   UNII table.

large amounts of data quickly. The most data intensive operation that a smart phone or tablet may perform is probably video conferencing. Video conferencing will likely only consume 3 Mbps or less bandwidth, which is more than within the limits of a 65 Mbps data connection. In light of this, it is important to keep in mind the target device when planning a wireless network. 802.11n access points can be configured with 20- or 40-MHz channels. Most mobile devices, particularly the most widely sold ones, are only capable of 20-MHz channels. 40-MHz channels are also called channel bonding. This is often represented with a +1 or –1. Figure 2.4 shows all possible channel-bonding scenarios for the 5.0–GHz channel.

Tablets and Smart Phones

Apple's overnight success certainly has had an impact on the medical community. The iPad is the tool of choice for many physicians during rounds and other highly mobile job functions. While wireless carts provided a level of mobility and power, in the end they are bulky and difficult to port around. These are the types of trade-offs that will always exist when engineering a solution. Tablets and smart phones have been a game changer simply because of the trade-offs that were made in engineering them. They are lightweight and have a long battery life relative to a laptop computer. On the other hand, mobile devices have less processing power and visual real estate. In addition, the wireless radios used on mobile devices must conserve power. This is accomplished by eliminating features like multiple spatial streams and 40-MHz-wide channels. These technologies allow wireless clients to transmit at 450-Mbps and faster. It is unlikely that a tablet or smart phone will have a radio configured to transmit at these higher data rates. Most will only transmit at 65 Mbps, which is only a minor improvement over the 802.11a/g maximum rate of 54-Mbps. Lower data rates affect the performance of all the clients connected to an AP. It takes more than four times as long for a client to transmit 1 MB at 65 Mbps than at 300 Mbps. It could also be said that the other clients connected to the AP must wait over four times as long. It is still possible to have a tablet or smart phone transmitting an HD video at 65 Mbps, just fewer of them. This in turn affects the maximum number of clients that the network can support.

Bonjour

With the launch of Mac OS X and the later success of iOS, Apple needed a way to connect all of these together. Enter Rendezvous, which of course later was renamed Bonjour. An avid Apple user may know this technology by its name. However, average users will simply know how your iPhone connects to your Apple TV. Bonjour uses a special type of IP address called multicast. Multiple devices can "join" a multicast address in order to efficiently receive communications from within a group. There are three types of IP communication: unicast, broadcast, and multicast. Unicast frames are between two endpoints. Broadcast frames are flooded out to all clients on the same subnet. Multicast frames are sent to all "members" of the multicast group. It should be no surprise that Bonjour uses multicast to create and group devices that communicate among each other. Multicast and broadcast frames can be very deadly to wireless LANs. As discussed earlier in the chapter, these types of frames are sent at the lowest Wi-Fi data rate. And since these are also flooded out to every port that is a part of an IP subnet, every AP on that subnet may be broadcasting this single frame. Fortunately, these frames cannot be routed. There is a field in an IP frame called the TTL or Time To Live. Every time the frame is routed between subnets this value is decremented. The TTL of a Bonjour multicast frame is one. Once the TTL of an IP frame reaches 0 it is dropped. There are gateways that vendors have introduced to overcome this limitation. It is important to keep in mind that these type of protocols need to be approached with caution in a large network. They were designed specifically for a home network. There are many uses that are valuable in the enterprise such as printing and streaming presentations from an iPad to a printer or projector, respectively. In order to accommodate this sort of functionality it would be necessary to have all these devices on the same subnet or implement a Bonjour gateway. There are other protocols that operate very similarly to Bonjour. Also, increasing the minimum basic rate will improve the efficiency of the WLAN. However, this must be taken into consideration during the site survey phase.

Wireless LAN architectures will continue to change and evolve as new problems emerge to solve. One thing that will never change is the process of designing a network. The starting point for wireless LAN design is identifying which devices and applications will be utilizing the wireless LAN. Also, it is important to understand the trade-offs introduced by the client devices, applications, and wireless LAN infrastructure. Finally, every wireless LAN design should have clear goals identified. Following these simple guiding principles will ensure success in delivering an effective design.

3

SITE SURVEY PROCESS

Wireless Site Survey Process

A properly executed site survey is the foundation of a fundamentally sound wireless design. The wireless site survey process encompasses many critical components to plan, execute, and design a robust hospital-grade wireless network. The authors of this book have witnessed a number of flawed wireless networks that either were installed with no survey whatsoever or were installed by persons who lacked the expertise to get the site survey process right. With the demand for robust hospital-grade applications such as voice over Wi-Fi (VoWiFi), real-time location services (RTLS), and real-time video it is essential to optimally design wireless networks with a well-executed site survey process.

This chapter will help you implement the survey process on your own or provide the pertinent information necessary to select a company to complete the survey for you. There are many companies that offer site survey services, but with varying skill levels. With this information you will be able to intelligently interview and hire a company to perform a proper survey.

Preparation

Preparation is the most important first step in designing a wireless network. It is essential to clearly define exactly what work has to be done for a successful survey before beginning the work. Often this will require work effort exceeding the actual time to complete the technical portion of the survey. The next section will cover components needed prior to the technical work effort.

The Statement of Work

The statement of work (SOW) is a formal document that defines the work effort, timelines, and deliverables a surveyor must execute to complete all requirements of the project. Producing the background information needed to complete the SOW will require diligence. Closely aligning with the customer contacts and performing interviews of as many stakeholders as possible will help factor in the necessary scope data to produce a statement of work.

The list below provides areas addressed in a SOW:

Purpose: What is the specific need for the wireless infrastructure? What applications and devices are going to be used on the network?

Scope of work: This describes the work that must be done in detail. It documents the exact nature of the work and specifies the wireless hardware and software involved.

Location: This describes where the work must be performed. What are all of the physical address locations involved? Where will the work be done?

Period of performance: This specifies the timeline for the project. This should include start and finish time, number of hours that can be billed per week or month, and anything else that relates to scheduling.

Deliverables schedule: This specifies when deliverables are due. When is the site survey documentation to be completed?

Applicable standards: This describes any industry-specific standards that need to be adhered to in fulfilling the contract. The hospital has several standards to factor in aside from the technical standards, including HIPAA and HiTech requirements.

Acceptance criteria: This specifies how to determine if the survey is deemed acceptable. What criteria will be used to determine this?

Special requirements: This specifies any special hardware or software, specialized workforce requirements, such as degrees or certifications, travel requirements, and anything else not covered in the contract specifics.

Payment schedule: Payment breakdown must be determined up front. Terms will be negotiated in this section. Will the payment for services be up front, paid by percentage of completion milestones, or after the survey acceptance is complete?

Once all of the requirements have been gathered and the statement of work is completed, a kick-off meeting with all of the primary stakeholders needs to be scheduled. Formulate a concise meeting agenda to keep the topics on task. This meeting will review the SOW and should be agreed upon by all parties prior to beginning the work.

Facility Blueprints

Building blueprints or some form of floor plan will be needed for the survey regardless of the chosen survey tool. We have found this to be one of the most challenging items to produce. Adobe PDF or AutoCAD DWG files are usually preferred but don't be surprised if you have to use a copy of a fire escape plan or actually have to produce a floor plan by hand. When resorting to creating prints by hand it is extremely important to scale the print appropriately. This is the golden rule as a floor plan out of scale could potentially cause an improper design. Microsoft Visio or AutoCAD software can assist in creating a floor plan. It is recommended to ask for plans up front and be sure to factor in extra hours if you must create the floor plans by hand.

Pre-Survey Walkthrough

A facility walkthrough will ensure there are no surprises prior to scope completion. This may cause extra travel expense but should be a consideration where a miscalculation of the scope and timing of the project could yield an unhappy customer and an underpaid surveyor. The walkthrough may also provide an opportunity to discuss requirements with stakeholders at the facility. Validate the in and out of scope areas and print accuracy. Hospitals are almost always under construction, making sure prints are accurate up front will save heartache later. This is also a good time to note challenging areas to cover and get a better idea of the material properties of the building. These facility details will be great preparation for the technical work and will increase the accuracy of the project timing. Almost all hospital environments will require physical access challenges that are better addressed up front than when busy capturing data. Keep in mind other factors that may cause complications. Access to restricted areas, including clean rooms, pharmacies, and mental health areas, may require escorts or permission up front. Operating rooms are always challenging areas to gain access to. Schedule these visits carefully during off-peak hours. Even then an emergency case will require rescheduling. Working as a wireless professional in the hospital will require flexibility. Understand that it will be nearly impossible to plan for every hurdle you may encounter inside the hospital environment. Planning for the unexpected will help keep project timelines on target.

Design Considerations

After you have interviewed the customer extensively and have gathered all of the information above, you will need to determine the RF baseline for the survey. It is important to understand how different wireless requirements will change the design and placement of access points. The explosion in number of medical devices has created a need to shift design methodology to analyzing the client density and client type. It would be extremely cost and time prohibitive to analyze every client device that could impact the initial wireless design. Be certain that once the network is operational the number of clients is guaranteed to diversify and grow. The best practice is to design to the most demanding client device use-case. This will almost always end up being voice or real-time video applications. These applications require the least network latency to function properly. This means the wireless design needs to be well executed to provide for these services. Voice has an industry standard signal requirement of −67 dBm with a −25 dBm signal-to-noise ratio (SNR). Using these baseline RF requirements will provide the design objective for placing access points. More details about voice over Wi-Fi will be in covered in Chapter 7.

The survey baseline will need to accommodate minimum RF signal needs as well as taking into account the geographic density of client devices. Nurse stations, board and conference rooms, cafeterias, and large waiting rooms may need a larger number of access points to appropriately accommodate user load.

If RTLS is a requirement, the placement of access points will need to be configured to facilitate device triangulation. This will change considerably the design methodology and placement of access points. We will discuss RTLS design considerations in detail in Chapter 8.

High-Capacity Design

As wireless has evolved it is no longer sufficient to simply design for basic RF coverage. The explosion of mobile devices has created the need to add capacity to the design consideration. For the last few years the authors of this book have designed wireless networks for maximum device connectivity. This essentially means installing significantly more access points with a lower transmission power so that RF cells are much more compact than traditional coverage designs. Dense AP deployments provide greater bandwidth because there are fewer devices connected to each access point. As Wi-Fi is a shared medium, less contention yields higher data throughput. This expanded capacity adds several design constraints. Co-channel and adjacent channel interference is one of the greatest obstacles to overcome. The 2.4-GHz range is especially challenging as there are only three non-overlapping channels. In a highly dense environment the AP power needs to be set very low in order to minimize interference. These types of interference can significantly degrade network throughput for poorly managed channel and power plans. To mitigate co-channel interference, planning appropriate channels and power will be necessary. Turning off legacy data rates below 18 MB will help shrink your RF cells and force clients to roam more aggressively. Legacy client devices may impede the ability to turn off the lower rates. Do some diligence to find and remove any 802.11b devices that may be on the network. If any are discovered make a case to leadership if necessary to have these clients replaced. There is a good chance these legacy devices will also not support the preferred encryption schema. It is best practice to turn off these lower rates to increase capacity by conserving airtime. Less client contention increases throughput resulting in better client performance.

Most wireless equipment manufacturers provide additional features to help manage high-density deployments. We will not go into the details of how each vendor implements these features but will illustrate the concepts below. These features should not be used as a substitute for a proper wireless design.

Airtime fairness—is designed to provide more airtime to faster clients by allowing dedicated airtime to different bands. This can help mitigate performance degradation of mixed client environments.

AP load balancing—this mechanism is intended to move clients from overloaded access points to nearby access points with less client load. This may be especially helpful for conference or waiting room areas where a high number of clients will occur on a regular basis.

Band steering—is a radio management mechanism to increase capacity and throughput by directing faster clients to a 5-GHz band. This results in increased airtime for faster clients and can improve client performance for the more contentious 2.4-GHz band.

Beam forming—is a signal processing technique used to control the directionality of the transmission and reception of radio waves. The full advantages of these features are yet to be fully realized; however, the concept could significantly increase RF efficiencies.

Role-based policy management—provides a mechanism to implement network policies based on the grouping of clients by their role in the organization. Grouping similar behavior profiles such as physician, resident, and guest users can be used to leverage the amount of bandwidth, the QoS, and the order of priority in which a client can communicate on the wireless medium. For example, guest access may have a rate limit of 1 MB/s and best effort quality cue where the physician who uses telerounding video equipment may have a video cue with top network priority. As wireless clients continue to grow exponentially, the need to intelligently manage who gets what access to the network will be a critical success factor.

Channel Planning

As discussed in Chapter 1, the Federal Communications Commission (FCC) provides the unlicensed spectrum available for Wi-Fi use. As this book is being written, the FCC is also evaluating the need for expanded spectral capacity to meet the demand of the multitude of

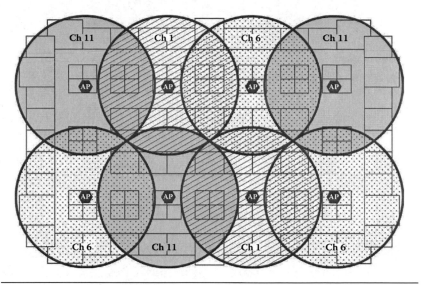

Figure 3.1 A 2.4-GHz channel plan.

mobile clients contending in the unlicensed spectrum. The indus-trial, scientific, and medical (ISM) band can be challenging due to the mere three non-overlapping channels. A wireless network must have a well-balanced signal overlap to meet the demands of seamless mobility; however, too much overlap in the same channel will create interference. Inexperienced engineers often misunderstand the signif-icant impact of APs interfering with each other. Figure 3.1 illustrates a proper ISM channel plan. Note that the overlap of access points for seamless mobility should be close to 20 percent.

Multifloor Designs

Hospitals are often very large multiple-floor facilities. Multiple floors will add to the complexity of your design. Minimizing floor-to-floor RF bleed and co- and adjacent-channel interference will take careful design. It is important to take measurements of floor-to-floor attenu-ation. This should be done by placing the AP in a desired location and recording RSSI and SNR measurements on the floors above and below. This may be done best with two people, one to watch out for patients and clinicians while configuring power-setting changes with the surveyor on the other floor. It may take several adjustments to find the desired RSSI on the adjacent floor. Access points that are stacked on top of each other can cause significant interference under the right conditions. Minimize interference variables by placing APs in a staggered manner. Optimal placement is preferably near the roaming boundary of the access points above.

Aesthetics

Depending on the customer and the type of facility aesthetics may play a major role in wireless design. The authors of this book have been required to explore this area at great lengths. It may be necessary to find creative ways to blend APs into the existing environment.

Custom painting (Figure 3.2) may be a way to blend devices into the surrounding area. If you must paint a device to match an existing area remember these important facts. In most cases painting a device

Figure 3.2   Custom painting.

Figure 3.3   A custom enclosure.

or antenna will void its warranty. If painting is required make sure to use a non-lead based paint that will not impede the RF properties of the equipment. Be sure to plan accordingly as custom painting will require extra time and money to complete.

Custom enclosures (Figure 3.3) may be a good way to protect devices from theft and damage. Make sure you evaluate RF qualities before choosing a design. Ceiling mount enclosures may be an alternative to plenum mounting making the device accessible without using a particulate tent. Enclosures may be expensive but may provide a clean appearance that is secure and easy to access.

Light patterns on a device may disturb patients. A flashing AP in a hospital room or a mental health area may cause patient dissatisfaction. Most manufacturers have options to disable the access point light patterns.

Augmenting Existing Designs

It is becoming increasingly rare to find hospitals without at least some wireless footprint. As a result it is important to understand design considerations for augmenting existing wireless hardware.

Upgrading Access Point Hardware   Wi-Fi technology has evolved very quickly over the last ten years. Legacy wireless networks do not have the capabilities to support the increased demand of applications in the hospital environment. As a result many organizations are making large capital investments to upgrade the wireless infrastructure. Migrating from an 802.11 a/b/g to 802.11n and soon 802.11ac requires completely new variables to overcome. The Multiple in Multiple out (MIMO) technology introduced in the 802.11n standard really was a game changer for wireless design. Before its introduction, an RF characteristic known as multipath caused signal disruption and retransmissions. Multipath occurs when a radio signal is split into multiple signals causing the receiving antenna to receive multiple copies of the same signal. The radio signal can be split by obstacles such as walls, windows, lead lining, and other objects. Signals may bounce off several objects before reaching the receiver causing delay. As radio signals are delayed they reach the receiving antenna at different times. Pre-802.11n performance can be significantly reduced by the delayed signals and retransmissions. Augmenting or replacing access points that were installed before the introduction of MIMO technologies will require a complete site survey process.

Cabling

Existing cabling may not support the gigabit speeds needed to support modern wireless access points. Distance of the cable is an important variable. Cables exceeding 100 meters may not provide sufficient power to access point hardware. Remember when upgrading from a 100-MB device to a gig-enabled device that additional pairs in the cable will need to be functioning. We have found a surprising number of cables during AP replacement that fell into this category. Do not make an assumption that a reuse of cabling is an option without sufficient testing.

Network Infrastructure

The wired infrastructure provides the backbone for the wireless network. Several critical elements should be validated to ensure a successful deployment, including port availability/bandwidth, power, WAN bandwidth, and IP address availability. If the customer cannot provide details on these elements they may need to be collected during the survey process. Remember a rule of thumb is to make sure if you have to collect this type of data on a network you do not administer to have a read-only account. This will reduce your liability in the event a problem surfaces while you are looking at the network.

Network Ports

Modern 802.11n access points require at least one gigabit-enabled port. Many manufacturers have an option to add an additional gigabit port depending on the number of radios and spatial streams. It may also be necessary to validate the uplink capacity of the switch to the head end to ensure enough bandwidth is available. Ensuring the proper number of ports are available and set aside for wireless prior to installation will save a big headache if you find out during the installation that additional port capacity is needed or that a new voice system was installed since you did the survey and the open ports are now in use.

Power Availability

Most access points need 802.11af Power over Ethernet (POE) ports to facilitate power. This will provide the standard maximum of 15.4 watts of power necessary for some access points. Some access point manufacturers will require more power. Running two Ethernet cables may be an option to power devices requiring up to 30 watts of power. 802.11at POE may be a consideration as the next generation of 802.11ac devices become more prevalent. It is important to look at the power specification on the switch as POE switches may only be able to fully power a percentage of the total number of ports simultaneously. Power capacity is documented by all switch manufacturers to help such calculations. In some cases adding additional power supplies will help facilitate more available power. Keep in mind the maximum distance of a CAT 5e/6 Ethernet run should not exceed 100 meters. Exceeding this length may not provide enough power for the device to function properly, causing network problems. Finally, a power injector may be a solution to provide power if a POE switch is not available. If this option is chosen a sufficient number of wall outlets must be available in each closet location.

Network Bandwidth

Depending on the architecture of your wireless design it may be necessary to factor in WAN bandwidth into the design consideration. If using centralized controller architecture the access points will transfer all traffic back to the controller infrastructure over the WAN. More details about the variables to consider and the design types are provided in Chapter 2. The important factor to consider is if there is enough bandwidth if WAN traversal is needed. For instance, we require a minimum circuit size to provide guest access. For smaller sites with low bandwidth circuits the addition of WLAN services may compromise bandwidth availability for production clinical traffic.

IP Address Availability

The need for IP addresses has grown exponentially with the rapid increase in number of mobile devices in hospitals. IP addresses are needed for client devices to communicate on the network. There are many considerations that need to be taken into account, including least time, subnet sizing, failover capacity, and broadcast size. This is discussed in Chapter 2. For the purpose of this chapter it is important to note that the scope of IP addresses will need to be larger than you would ever predict.

Survey Equipment

A variety of specialized equipment is necessary to provide a proper site survey. The Boy Scouts of America have a motto to "be prepared." This approach in the field is highly recommended. We will discuss the equipment needed to complete a survey in this section. We have used a large variety of hardware and software before finding just the right combination of equipment that works most efficiently.

Onsite surveys will require an apparatus to mount the chosen wireless access point. This comprises a telescoping pole, mobile power source, and antenna mounts to emulate the exact hardware of the design, including antenna orientation and height. The top unit should be adjustable horizontally and vertically to emulate wall and ceiling mounting. The pole should have the ability to reach a minimum of

12 to 15 feet. Choose a battery pack that can last an entire day in the field and matches the power needs of the hardware you are using. If the AP hardware needs more power than standard 802.11af it will be essential to ensure your battery pack is capable.

Use a mobile unit with wheels so that it can be moved easily throughout the hospital facility. These wheeled units often also have a smaller footprint than a tripod for the busy hospital environment and can be moved quickly in the event you must move out of the way of patient care. Many manufacturers sell predesigned "survey on a stick" kits (Figure 3.4) but it is not uncommon to build a custom apparatus. We once used a camera tripod that had seen heavy use. At one point the tripod finally surrendered to age and fell over. Fortunately no one was around at the time but it could have been a much worse situation

Figure 3.4 Survey on a stick.

if a patient or clinician was nearby. The moral of the story is you want a sturdy cart that can withstand the rugged hospital environment. If air travel is required, your kit should be designed to break down for easy transport. Pelican cases can be customized with foam and provide an excellent vessel to transport gear.

Mentally combining our Boy Scout motto with episodes of MacGyver will help you keep in mind that you will require a mix of miscellaneous items that you may not use often but in a bind may make or break your timelines for a survey. Carrying multiple access points with both internal and external antenna models is recommended. Building challenges may require different mounting options that may be better suited for one form factor over the other. Depending on the job size a second mobile battery unit may need to be added. Other key items may include a basic toolset that includes extra mounting hardware, duct tape, zip ties, a camera, safety marker, appropriate cables, and of course dental floss and a gum wrapper.

The next section will cover the client form factor used to measure the RF data.

Form Factor

There is much industry debate about what client to use to capture data. It is important to use a consistent setup while doing a survey. Using different clients with different chipsets may change the readings recorded, thus creating inconsistencies in a design. Here are factors to consider when choosing a measurement form factor. A portable computer is needed for the survey process. At some point you will be carrying a device even if you use a cart. Many areas in the hospital may not have enough space to maneuver a cart. Size and weight should be a consideration for your physical well-being. Extend battery life by choosing a device that supports high-capacity batteries, and having extra battery bays will help keep you going for the long hours necessary. We have had great success using high-end tablet computers. Make sure the form factor you choose has an operating system compatible with the site survey software you plan to use. Configuring the form factor to emulate the baseline client device is recommended. Figure 3.5 shows an OptiView XG network tablet.

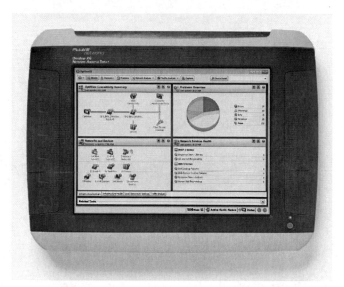

Figure 3.5 OptiView XG network tablet. (Photo provided by Fluke Networks.)

This is a very high-end tablet that hosts a number of mandatory tools on one easy-to-use platform.

Site Survey Design Software

Site survey software is used to plan and design a successful RF design. Most tools go beyond simple RF coverage and take into account throughput and client statistics to help achieve a high-capacity design to provide a robust client experience. Choose software that provides a solution that accommodates all of the different survey types discussed later in this chapter. Most manufacturers will provide a short trial demonstration so you can see what interface works best. It is highly recommended to also attend the formalized training made available by most software manufacturers to ensure proper design.

Regardless of what software suite you choose, the fundamentals of using site survey software are the same. A facility blueprint is imported into the software. Accurate scaling is the golden rule. Most software has the ability to choose a number of wireless hardware including access points and antennas. Some even have the azimuth and elevation parameters of common hardware elements built in. The software allows placement of equipment on the floor plan to prepare

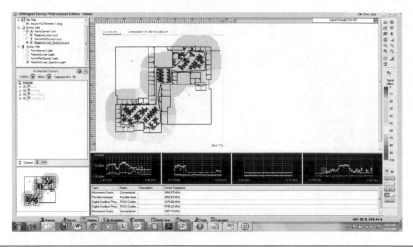

Figure 3.6   Survey Pro. (Photo provided by Fluke Networks.)

for onsite measurements. Figure 3.6 is a capture of a sophisticated site survey tool. Note that this facility has excellent coverage with an RSSI between –50 and –60 dBm. This software is also displaying the spectrum analysis data along with survey information. This is a great combination of important data measurements and could be a real timesaver by completing the spectral analysis and the design together.

Spectrum Analyzer

Spectrum analyzers are required for a proper onsite survey. Spectral analysis will identify sources of RF interference that may cause problems with the intended Wi-Fi design. There are two primary portable options to choose from: a dedicated handheld analyzer unit and software loaded on the chosen form factor. We use two tablet computers, one dedicated to spectrum analysis and one to capture readings on both the 2.4-GHz and 5-GHz bands. Using multiple computers minimizes the amount of laps that are needed, thus yielding big time savings. More will be discussed about spectrum analyzers in Chapter 10.

Survey Types

There are three types of site surveys, including predictive, passive, and active surveys. There are advantages and disadvantages to each method when designing a wireless network.

Predictive Survey

A predictive site survey is the process of wireless design by utilizing simulation software. A virtual model of the building is created containing the material properties of the facility. Once the model of each floor is complete a model of the AP and antenna can be placed to visualize RF coverage. Many survey software vendors provide building modeling. This is done by tracing on blueprints of the buildings the physical attributes of each floor. The software then assigns generic attenuation properties to each of the material properties based on the assigned building properties. Design accuracy is solely dependent on the accuracy of the model. In a hospital environment it can be especially tricky to build an accurate model due to construction changes. We have used predictive modeling to simulate the optimal AP placement only to find out the new management wing was once a radiology area with lead-lined walls. Although the material properties appeared to be drywall this was not the case. This example represents the primary reason why predictive methods should not be used in the hospital environment as the only method to build a design. Without an onsite analysis of the interference mediums and other RF qualities, a hospital can be sure to provide surprises that may jeopardize the functionality of the WLAN.

Although a predictive survey is a successful methodology to get a start on a wireless design, it should never be used as a substitute for an onsite analysis. Predictive site surveys are a guide used before performing one of the other survey methods. Other industries may be able to use the predictive survey as a way to save time and cost when designing facilities that allow for a higher margin of error. If working for a customer that absolutely does not have the investment to do a full site survey you should explain up front in detail the limitations and possible repercussions of a predictive survey.

Passive Survey

A passive site survey is a process of detecting active access points, signal strength, channels used, and noise level. This method includes using site survey software and placing the adapters in listening mode. This method provides an understanding of the RF characteristics. We highly recommend this method be used during the design phase even for just one pass through the facility. This is a good way to detect interference mediums within the space you are designing. Identify rogue or other wireless devices in the facility and provide a list in the survey report. It may be necessary to address these devices prior to install. Passive survey is also a good method to use for post-survey validation. We will discuss this subject in more detail later in the chapter.

Active Survey

Active site survey is the process of associating to the "AP on a stick" to take measurements of the AP connectivity and performance. The key differentiator to this method is the Wi-Fi adapter receives and sends packets. Active surveys provide metrics like throughput, round trip time, packet loss, packet delay, retransmission, and other useful metrics. Active survey can be done using two different methods by associating to the BSSID is the Basic Service Set Identifier or Service Set Identifier (SSID).

The BSSID method uses the client Wi-Fi adapter to associate to a single access point. This method will provide the actual client performance metrics and provide the full coverage pattern of the AP.

The SSID method associates to several access points and can provide additional metrics like roaming but will leave out potential critical data pertaining to the actual coverage area of a single AP as the client will roam from AP to AP. This is also an excellent method for post-deployment validations.

Survey Techniques

Both passive and active surveys will require an engineer onsite at the facility to take measurements. It is important to understand where you are physically located on your floor plan to ensure accuracy. Stairwells and elevators can be good points of reference to ensure you know exactly where you are on the floor in relation to your blueprint. As you move through the facility record collection points by marking your location on the print. Keep a consistent walking pace during data collection. The more data points collected the better. It is good practice to take data points in as many areas as possible. Be sure to take measurements on all sides of material obstructions. In the hospital environment this can be very challenging especially in areas that contain patients around the clock. Remember most wireless transactions will occur in the patient rooms. It is a critical success factor to take measurements in these areas. It may be necessary to work with clinicians to gain access to these patient areas. They can usually help get permission from patients before entering. Surgical areas may also prove to be a difficult area to access. Work closely with your contacts to schedule the best off-peak times to be in those areas.

Consider the best time to complete the survey. Taking measurements in the evening when the building is empty may not yield the best results once all of the employees are present. This is especially important in high density areas such as waiting, conference, or boardrooms. Keep in mind workflow. It may be common for a nurse to be in a room with the door closed; therefore, you should take measurements with the door closed. Putting in this extra work during the survey will give you the clearest picture of the RF attenuation.

Site Survey Report

Most site survey software provides exportable templates to deliver a professional survey report for the customer. There are several outputs the site survey report should contain.

- A design summary provides details of the baseline RF design. This will include the RF bands used, the wireless manufacture, and high level architecture for the design.
- The site survey process documents the steps taken during the site survey. This is a good way to ensure the agreed-upon methods were utilized during the design.
- A bill of material (BOM) is a detailed listing of the make, model, and number of all of the hardware necessary to install the design.
- AP placement and configuration provides blueprints that clearly mark the AP locations to ensure proper installation. Document the channel and power plan along with the antenna and access point orientation.
- Heat maps provide blueprints illustrating the recorded RF coverage map. These maps will ensure the baseline RF coverage is met in each area.

Post-Validation Survey

Once all of the hardware is installed in accordance with the site survey report, documenting the post-validation survey is the last step of the survey process. This step verifies the intended design is appropriately installed. Using both active and passive survey techniques can be used to validate during this phase. This process may result in configuration and optimization changes to the system. Ensure all of the AP devices are installed in the correct location with proper orientation. Validate the blueprint placement matches the actual installed placement of the access point. When installing APs, unanticipated obstructions above the ceiling may be discovered. It may be necessary to move the device a short distance to overcome mounting challenges, which may adversely impact your intended design. Work closely with cable vendors to ensure this is not a surprise.

A baseline of the RF should be recorded to ensure the agreed-upon minimum RSSI metric is realized. If –67 dBm is the baseline you should have at least this coverage everywhere in the facility. Ensure that co- and adjacent-channel interference is minimized by validating the channel and power plan.

An often overlooked but critical item for the post-validation survey is throughput and performance testing. Many vendors offer software to help gauge performance. Even if baseline RF coverage and channel plans are perfect, you may not realize other variables could cause issues. We were called to a facility experiencing horrible throughput and overall performance. An RF validation was completed and the wireless signal looked very good. After a few performance tests it was evident there was a non-RF-related issue. The issue was badly configured switch ports settings, limiting the AP link speed to 10 MB rather than 1 GB. Once resolved, the anticipated wireless performance was realized.

The hospital environment is a very hostile environment for the propagation of radio frequencies. This environment has unique physical construction properties that require a "boots on the ground" approach to ensure a proper design. Following a well-executed site survey process will ensure the wireless design is done properly the first time. Proper preparation, careful design considerations, the right tools, survey methodology, and validation will result in a successful robust hospital wireless design.

4

WIRELESS SECURITY WI-FI

About Information Security and Wireless Networking

Information security is the delicate balance of three key factors, which are often referred to as the three tenets of information security. They are confidentiality, availability, and integrity. Without the careful balance of these three items, any given security control measure could end up either too restrictive or so easy to use that security is almost nonexistent. Understanding these three key items will help balance any security process to ensure it is usable and secure. This information will help when we start looking at all the risk and threats that face Wi-Fi communications.

Confidentiality

Attacks on the confidentiality of information relate to the theft or unauthorized viewing of data. This can happen in many ways, such as the interception of data while in transit or simply the theft of equipment that the data may reside on. The goal of compromising confidentiality is to obtain proprietary information, user credentials, trade secrets, healthcare records, or any other type of sensitive information.

Attacks on the confidentiality of wireless transmissions are created by the simple act of analyzing a signal traveling through the air. All wireless signals traveling through the air are susceptible to analysis. This means there is no way to have total confidentially since you can still see a signal and subsequently record it. The use of encryption can help reduce this risk to an acceptable level.

Availability

Availability allows legitimate users access to confidential information after they have been properly authenticated. When availability is compromised, the access is denied for legitimate users because of malicious activity such as the denial of service (DoS) attack.

Receiving an RF signal is not always possible especially if someone wants to prevent you from doing so. Using a signal jammer to jam an RF signal is a sizable problem that has faced national governments for many years. Looking into the availability of RF local area networks (LANS), we notice that carrying out a DoS attack is extremely easy to accomplish.

Integrity

Integrity involves preventing the unauthorized modification of information either while it is in transit or while it is being stored electronically or via some other medium. To protect the integrity of information a validation technique needs to be employed. This technique can be in the form of a checksum, integrity check, or digital signature.

Wireless networks are intended to function in an unimpaired manner, free from deliberate or inadvertent manipulation of the system. If integrity were not upheld it would be possible for an attacker to substitute fake data. This could trick the receiving party into thinking a confidential exchange of data is taking place when in fact it is the exact opposite.

Wireless Security Risks and Threats

Denial of Service

DoS is a type of attack that renders a network device or entire network unable to communicate. Hackers have found that certain crafted packets will make a network device unresponsive, reboot, or lockup. They have used this technique to shut down high-traffic networks and websites. They have also used this attack to reboot network equipment in an attempt to pass traffic through the device as it is booting up. This is done to try to circumvent any policies set up on the device to protect it or devices behind it. The DoS threat can also adversely affect the availability of a network or network device.

Wireless DoS attacks can be achieved with small signal jammers. Finding signal jammers is not as difficult as one might think. Some modern wireless test equipment can perform jamming. This is not the tool's intended purpose although it is commonly used for this. Jamming is possible because the government regulates the amount of power allowed on a wireless network. Only very small amounts of power can be used on wireless local area networks. This means that it is not hard to overpower an existing commercial wireless device with a homemade one capable of superseding the power output regulated by the government.

Another DoS threat relating to local area networks in particular is the poor structure of management frames. These frames allow anyone who can analyze the wireless signals to perform a DoS attack by replaying certain management frames. Most of these attacks are layer two-frame attacks. These attacks try to spoof management traffic informing the client that they are no longer allowed to stay connected to the network. Chapter 13 discusses these attacks in more detail.

Malicious Code

Malicious code can infect and corrupt network devices. Malicious code comes in many forms, including viruses, worms, and Trojan horses. These three main forms of malicious code are often confused. Because of this, many people use these terms interchangeably. In this section, we will look at each of these and identify what classifies them into each of the three groups. Viruses infect devices and do not have the ability to replicate or spread outside the infected system on its own. Once a virus infects a machine, it can only replicate inside the infected machine. This means all threats from viruses stem from receiving infection. The threat of worms is much higher because they can spread across the enterprise and out to the Internet, infecting multiple devices. In the past few years humans have started to see worms that propagate across the entire world. The last malicious code threat we are going to discuss is the Trojan horse. This threat comes from installing or running programs that can execute code that may have malicious content. The program or application the end user thinks he or she is running may actually run without issue, although the malicious code is also running in the background just as if the user directly executed the malicious code.

Malicious code relating to wireless has to do with new viruses that can affect the many new types of wireless end devices such as PDA units, smart phones, PDA phones, laptops, and much more. Wireless viruses have just started to appear that either target these device types directly or take advantage of wireless communications. There have been viruses that spread via bluetooth that actually spread like a biological agent would because you have to be within close physical proximity. Even with this threat just starting to develop, many forms of wireless malicious code have already appeared. Some of this code has enough intelligence to find and utilize a variety of available wireless technologies on a device to spread even further.

Social Engineering

Social engineering is often called low-tech hacking. It involves someone using the weakness of humans and corporate policies to achieve access to resources. Social engineering would be best defined as tricking or manipulating a person into thinking the party on the phone is allowed access to information, which they are not. The threat of social engineering has been around for quite some time. Some of the most well-known computer hackers used this type of attack to get information. The real threat is the skill level involved. One does not need to be computer savvy or a technical genius to perform this type of attack. A number of measures can be taken to prevent this type of attack. First, make sure that a policy is in place in your organization about sensitive information and phone usage. Make sure that not anyone can call and reset someone's password. Create a helpdesk identification process to authenticate callers to the helpdesk operators.

Signal Analysis

Signal analysis is viewing or recording a signal, or eavesdropping. There are many valid reasons for an authorized party to perform an analysis, although with wireless signals traveling through the air unauthorized analysis is a persistent threat. All RF signals are prone to eavesdropping because the signal travels across the air. This means anyone within the range of the signal's path can hear the signal and perform analysis on it. One of the only protections available to deter unauthorized signal analysis is to apply a common confidentiality control such as encryption. The risk of analysis on an RF signal is an inherent risk that cannot be avoided or prevented. The only option is to mitigate the risk with some type of confidentiality control thus deterring the unauthorized party.

Spoofing

Spoofing is the act of impersonating an authorized client, device, or user to gain access to a resource that is protected by some form of authentication or authorization. When spoofing occurs in wireless networks, it primarily involves an attacker setting up a fake access point in order to get a valid client to pass authentication information to an attacker. Another way attackers spoof is by performing a man in the middle attack. In this scenario, an attacker gets between a client and the network. This can be accomplished by spoofing a valid access point or by hijacking a session. Once this is complete, the attacker uses the authentication information provided by the client and forwards it to the network as if it came from the attacker originally.

Rogue Access Points

Rogue access points pose a major threat to any organization. This is because of the high availability and the limited security features of off-the-shelf access points. If a company does not approach the wireless LAN concept quickly enough, frustrated employees will take it upon themselves to start the process. When this happens, employees often put in wireless systems of their own. Even with most current access points supporting advanced security standards, the default configuration of an out-of-the-box access point is set to the least secure method. This has created a real threat because now a user can easily bring in a rogue access point, plug it in, and put the entire network at risk. The knowledge level required to install an off-the-shelf access point has almost become plug and play today. This means more and more people have the ability to place rogue access points. However, they usually lack the ability to secure these devices or even understand the risk they are posing to the company.

Most rogue access points come from employees, although as we will learn later there are cases where an attacker tries to set one up for easy return access. This was not a big issue until recently when the price of 802.11b access points fell well below one hundred dollars. In order to do this an attacker would need physical access and a network port. If a hacker wanted access badly enough, spending one hundred dollars for it would be a conceivable expense.

With companies investing in stronger security mechanisms it would be a shame to have an incident where an attacker gained access through an unsecured rogue access point. Because of the threats associated with rogue access points many companies have started to put controls in place to increase awareness and prevent their deployment. Many companies that entered the newly formed wireless security market adapted and created tools to detect rogue access points. Some companies have handled rogue access points by creating policies about wireless usage and imposing strict penalties for rogue access placement. Others have taken a second route and invested in Wireless Intrusion Detection Systems (WIDS) software.

Wireless Hacking and Hackers

Inherently, RF has many potential threats, such as interception, signal jamming, and signal spoofing. Because RF travels through the air, picking up the signal is as easy as being in the vicinity of the radio waves with the right hardware. Spectrum analyzers can detect radio transmissions that show the user the signal frequency. Depending on that frequency, an attacker may be able to identify the transmission right away. Most RF frequencies in the spectrum are reserved for specific uses. Once you are able to find a signal and map it to a reserved spectrum, you know who is transmitting it and in some cases why.

Getting more in tune with the majority of RF threats, we look at today's RF LANs. These, of course, have the same threats as all RF signals, although they do add a new dimension stemming from mass use and scrutiny. Just like cell phones, the more people that use them the more time people spend looking at how they work and what they can do to defeat any security that exists. This has been seen over the past few years as security flaws have emerged as a large number of users have started deploying wireless networks. Most wireless network setups are capable of working right out of the box. This has led more and more nontechnical people to deploy them. When you set up a wireless LAN right out of the box the default configurations are usually the most insecure ones.

Motives of Wireless Hackers   U.S. laws and law enforcement have taught us that a crime is rarely committed without a motive. Therefore, if someone spends the time to compromise an RF signal there always is a motive. Some of these motives can be as harmless as wanting an Internet connection to send a loved one an e-mail or as terrible as committing an act of terrorism against a nation or government. To understand why someone would try to compromise an RF signal let us look at some of the most well-known motives, such as to get a free Internet connection, commit fraud, steal sensitive information, perform industrial or foreign espionage, and—the worst of them all— terrorism. After understanding what motive or motives an attacker may have, you can better understand how much security you should apply to your RF signal. If your company deals with financial information, you probably are more at risk from an attacker then a small doll shop. Knowing who might attack you and why can ensure that the correct risk reducing actions are taken.

War Driving   After more and more people realized that out-of-the-box wireless LANs were generally set up in the most insecure mode by default, people started to exploit them. This new fad of identifying and categorizing wireless networks based on their security level has been termed war driving (see Figure 4.1). War drivers use equipment and software to identify wireless networks. War driving allows attackers to understand the security associated with any wireless network they happen to pass by. This equipment not only allows the war driver to identify, locate, and categorize wireless networks, it also allows them to upload their results to a website where their friends and everyone else who has access will be able to see exactly where these unsecured wireless networks are located.

This has even become so advanced that the use of GPS has been incorporated to give other people the exact locations of these insecure networks. Anyone can simply go online and get a map to the exact location of an insecure network identified by a war driver. This

Figure 4.1 A war driver.

culture has taken on its own members who are not, stereotypically, of the classic teenage computer cracker or hacker. Some older, wealthier people have begun to war drive. Some individuals have hacked their in-dash GPS into a war driving display. Others have mounted fixed wireless antennas onto their vehicles. Figure 4.1 shows a war driver using a Cadillac.

A new type of war driving called war flying has emerged. As the name implies, war flying is the act of scanning wireless networks from a plane, which has made it possible to cover a large distance very quickly. Just like flying in the United States military has evolved so has the concept of war flying. Remote-controlled drone aircraft are now used for war flying. There are many websites now that provide build-your-own remote-controlled aircraft with small wireless scanning tools for war flying and wireless hacking.

A more athletic approach involves war walking instead of war driving. In this concept, a war walker strolls down the street with a laptop either in a bag or out in the open. War walkers use the same tools and equipment as war drivers to identify insecure networks. This has gained in popularity as the prime profile for a computer hacker is a teenager who most likely does not have a car or is not old enough to drive. Consumer industries have even gotten on this bandwagon by producing tools to find wireless networks. There are even devices that connect to your keychain that will beep or light up when wireless networks are detected. This has turned the act of war walking into an event that can be performed without effort during any activity that requires movement. This means a war walker can perform this malicious activity while he or she gets milk for mom.

Tracking War Drivers   How would someone track down a war driver? The Federal Bureau of Investigation has had several public cases of arresting criminals who have used wireless networks to compromise retail store networks. They were tracked down so we know it is possible, so let us learn how.

Once the investigation starts, a forensic team arrives on site and dumps the configuration and stored memory of all network devices and servers that were affected. Once this data has been properly removed, in accordance with the chain of custody, it is examined at a lab. This examination is a time-intensive process, so much so that it can make many incidents considered not financially worth the effort. Many cases are too small to warrant the massive effort needed to investigate them.

After the lab results are examined, we can see where the perpetrator first entered the network. Because this happens on a switch connected to an access point, we can determine that they came in over the airwaves. Once this information is identified, the wireless network card's MAC address can be determined. This address is hard-coded onto the card by the vendor and is regulated in a sense, which makes it globally unique. Some clever hackers have the ability to change the card's MAC but, as time has shown. many do not take the time to do this.

After looking at how wireless war drivers can be tracked down, we get to a more important point about wireless devices. All bidirectional communication wireless devices emit radio waves, so in a sense, all wireless devices can be tracked in one form or another. Next time you see some amazing new RF technology remember the statement above. No matter what a vendor will tell you about their technology, any bidirectional communicating wireless device can be tracked.

The Hacking Process

Hacking has existed for a long time. Wireless networking has provided a new way into networks and resulted in some changes in how hacking is carried out. No longer is a network targeted, then examined and attacked. Today's wireless war driving has shown that an attacker will now find a network first then plan an attack and return to attempt to break into the network. War driving can be thought of as merely checking to see if any windows are left open or if the back door is rarely locked. Once an easy way in is found the hacker will go back and start the normal process.

Information Gathering   Step one to understanding a wireless hacker is understanding how they find a target as well as the information online about the target. This step is called information gathering. A wireless hacker will most likely find a target by war driving. We should all hope this is how every hacker finds us because in this scenario, an attacker is not looking for something specific nor are they targeting a company directly. They just happen to drive by and want to test the security or are just looking for a free Internet connection. Earlier we described why war drivers do what they do and how they go about doing it. Now we will expound on this by looking at the next steps after a wireless network is detected. When an attacker war drives a network without advance planning, this actual gives the company the upper hand. This is because the attacker's level of knowledge about the company is very limited. This upper hand will not last long if the attacker decides to plan an attack on this network and uses the common process we are about to discuss.

After an attacker has decided to plan an attack, information gathering is the next step. To gather information an attacker uses the biggest source of data ever put together, the Internet. Most people are unaware of the Internet's capability to provide information about companies or people themselves. Have you ever googled yourself? Try it one day with your first name and last name in quotes like this, "Aaron E. Earle". What comes back might amaze you. An attacker will do this with a company's name and immediately they will find a slew of information about the company.

Some of the information a hacker would be interested in is traceable because of activities your company may be engaged in without realizing it. A common item hackers look for is IT people from the company trying to get help about products or services. If someone at the target company needed some help with a product, one of the steps might be to try a message board for the answer. Most likely, they will leave an e-mail address or in some cases a disclosure message may be inserted automatically. This message most likely has the company name inside it in addition to the legal disclaimer.

If an attacker is able to find this information, he or she can assess what types of networking products a company is using. In this case, the most relevant information would be e-mails to message boards about the company's wireless network. The attacker may continue to gather useful information by using other avenues of information gathering, such as public records. Going to the Security and Exchange Commission's website would be an example of public records. This website has specific information on recent mergers and acquisitions, which could serve as another avenue of attack. Using other government information sources you can find information about a company, the names of management personnel, and its employees. If two companies are linked together through a search engine this might indicate that they are close partners. If this is true, there might be a network connection as well between them. If a hacker found that the partner of their target has a weak network security system in place, they might use it as a backdoor into the target's network.

Another type of information-gathering technique comes from domain registration information. When companies register their domain names, they must include information about themselves. This information is publicly available and can serve as a means of social engineering. There are a number of tools to perform this type of information-gathering attack. The most common are whois, nslookup, dig, and some others.

Another not so widely known or commonly used form of information gathering is the various business information websites that exist. One example is Dun and Bradstreet, D and B for short. This is a company that creates, tracks, and monitors business credit. For less than a hundred dollars, a hacker can pull a full credit report on a company, including details about partners, creditors, or any legal trouble the target may be experiencing. What an attacker can do with this information is find out the business owner's name as well as key senior members who are listed on the credit report. Once an attacker has this information, he or she could perform social engineering attacks on helpdesk personnel to reset passwords. You would be surprised how many processes can be circumvented when the right name is dropped.

Enumeration   Enumeration, foot printing, and scanning are all addressed in this section, and defined hereafter as enumeration. After an attacker has finished gathering information about the target company the next step is called enumeration. In enumeration, the attacker goes looking for everything he can connect to and tries to understand what type of products they are, what company made them, and what software or firmware versions they are running. This is the step where war driving comes into play. A war driver is in the enumeration stage when they scan a wireless network. A war driver may have skipped the information gathering stage altogether and gone right into the enumeration stage just by war driving an area and not a selected target. They may often find a network by war driving and then go back to information gathering before actually breaching the network.

Looking at the enumeration stage, we find another interesting step, switch or repeat. In the enumeration stage, you may only be able to enumerate so far until you are required to move to the next step of compromise. If an attacker has exhausted his enumeration capabilities, he may have to compromise a device in order to gain access into another portion of the network. At that time, the attacker may need to revert to the enumeration stage in order to test all the new devices that he or she can connect with now that a certain network device has been compromised. This is very relevant in wireless network hacking. An attacker who can enumerate your wireless network can only go so far, after which he or she must compromise the wireless network in order to gain access to other devices on the wired network. Once access has been gained the enumeration will need to start over on the newly discovered wired devices.

There are many tools used in enumeration, most of which are some type of scanner. Scanning wireless networks is a small subset of what a scanner can do. The type of scanner that is most widely known is the port scanner, which scans for open transmission control protocol (TCP) or user datagram protocol (UDP) ports. After identifying these open ports, an attacker can use another scanner to finger- or footprint the device by requesting and examining the information sent from these open ports. All operating systems (OSs) communicate across these open ports. How they do so is unique to the OS manufacturer, allowing an educated person or tool to easily identify the respective OS type. A scanner with the ability to finger- or footprint a network can examine the traffic responses from a device and understand what

OS it is by examining the traffic. This is because certain operating systems handle TCP\IP in different ways and create certain packet formats that give away the OS type.

Another type of scanner is one that uses Internet control message protocol (ICMP) and simple network management protocol (SNMP) to map networks and networking devices. This type of scanner is used to create a network map so the attacker can understand the layout of the companies' network. This scanner is sometimes included with a port scanner and needs to be run before a port scan in order to know what device to port scan.

The last type of scanner we describe is the security vulnerability scanner. This type of scanner can test network devices for a large number of known exploits and vulnerabilities. This is usually performed in a very short amount of time. This is the last step in enumeration; once this is done, all that is left is to actually carry out the attack that the tools identified.

Compromise The attacker will then attack the wireless network or another network device to gain some type of administrative access. Once this access is achieved the device is considered compromised. If an attacker can compromise a wireless access point, they can insert themselves inside the network and go after the next network target. Most hacking attempts drive towards the goal of compromising something on the inside of the network as that opens the door to the most information and access. Hacking into an access point achieves this without the need to bypass a firewall or security edge architecture and that is why access points and wireless networks are often attacked.

In order to compromise a device an exploit needs to be created or found. If one was to be found it would likely have to be compiled and executed. In order to find these exploits you need to know where to look; these exploits are available in underground websites or Internet relay chat (IRC) channels. Most of the time after an exploit has been out in public view for a considerable amount of time a tool may appear that allows an attacker to point, click, and compromise.

Expanding Privileges and Accessibility This step is very important in the hacking process. When we look at hacking from a wireless perspective we need to understand in order to get to this stage, a compromise of the wireless network has to have already taken place. If a wireless network is compromised then there is little else to do in relation to it. The attacker would most likely move on the next network target. This stage of the process has more weight in relation to servers and other network devices where access is achieved although it does not provide fully unrestricted access. The process of expanding privileges has to do with attacking a device as a restricted user and taking advantage of flaws to elevate the privilege level to unrestricted.

The next portion of this section has to do with accessibility or being able either to get back into a network or to find an easier way to control a compromised device. When we look at getting back into a network, we start to think about backdoors. There are many different ways to set up backdoors. Looking at wireless backdoors, we will see that an attacker could set up multiple backdoors in multiple places. Before we go over the many types of backdoors relating to wireless we need to touch on the other portion of this process, making compromised machines easier to get into. In setting up backdoors, we will see one example where a desktop or laptop computer needs to be compromised before the backdoor can be set up. In allowing easy accessibility to compromised devices, we need to look at how this can be achieved. Using tools such as virtual network computing (VNC) and network command line tool (NCAT) devices can allow an attacker to receive remote shells. Once an attacker gets a remote shell, they can install or activate some type of remote management software. This software will allow for remote graphical user interface operating system (GUI OS) manipulation. Looking at how to go about installing this is outside of the scope of this book. Just be aware that it is part of the process and needs to be done in order to place some types of backdoors.

Getting back to backdoors, no pun intended, we see an interesting backdoor example. This has to do with dual homed computers, such as laptops. If an attacker can find their way into a company laptop, they might be able to find one where the network connection and a wireless connection are both active. With almost all of today's laptops shipping with wireless network cards the odds of finding this is very likely. If an attacker can find this and compromise it, her or she can set up the wireless as AD-HOC and set up routing into the network. Now when the attacker connects back into the network, no matter what level of wireless security is present, the attacker can circumvent it. In order to get access to the laptop originally the attacker may attempt to hack into it at a hotspot or steal it temporarily and return it before anyone finds out. The most likely situation would be where the attacker hacks into a network through some misconfiguration and attacks the laptop in case the misconfiguration was detected and fixed.

The next backdoor is one of the most common and, luckily for the attacker, in most cases this backdoor already exists. This backdoor is rogue access points; they are not only used by employees who want wireless but by attackers who stumble across them. Most of the time these access points are deployed right out of the box with no security and using default settings. This makes the network an especially easy target for an attacker. The main reason that rogue access poses a threat and subsequent risk is that an attacker can use them as easy already set up backdoors into a corporate network.

There is one other type of backdoor that is very similar to the rogue access points we already talked about, except this one is placed by the attacker. If an attacker can gain access into the physical building, he or she can place a rogue access point. This is common in hospitals and healthcare-related places where the public has free access. In this scenario, an attacker purchases an inexpensive access point and finds a spot to hide it. The attacker can then mount his attack from a distance away from the physical security the building may have. With a high-gain antenna connected to a laptop an attacker could be located far away from any physical security force's line of sight.

Cleaning Up the Trails   This is the last step in the hacking process, where we learn what hackers do to clean up after themselves. If companies were to catch on to the existence of an attacker, it would be through log files, event correlation, and advance security tools such as intrusion detection system (IDS) and intrusion protection system (IPS) systems. To clean up after himself, the attacker purges logs and performs other housekeeping techniques to hide his existence. This step is required in the hacking process to prevent a company from figuring out an attacker has broken in. If this step were not performed, a large amount of evidence would remain as to how the attacker got in, where they went, and what they accessed.

When we explore this section in relation to wireless most of the same ideas and techniques apply. A wireless attacker would try to cover his tracks by changing the access point's log file. This would have to extend on to other network devices as well. Any servers the hacker attempted to access would have log files indicating the attempt. All of these devices would need to have their logs cleaned up after they were compromised. To make things even trickier is the existence of a syslog server. If a syslog server were present, all devices would send their logs to it. This means the attacker would have to attack the syslog server as well if he wanted to clean all traces of his intrusion.

One of the misconceptions is how most hackers are found out. What some attackers tend not to realize is if they attack a box and are unable to compromise it their attack will be logged. That log will remain for some time on the device. Some of the ways that the log would be removed are listed below.

- Someone goes through the log file and manually erases it.
- The attacker compromises the device and changes the log.
- The log has a specific setting that allows it to be overwritten after it has taken up a specified amount of disk space.

An attacker has many tools that have been developed over the years to remove specific entries or repopulate log files with bogus entries. These tools are commonly used by hackers to cover their tracks once they have compromised the device. They have different methods. Some tools actually change log events. They would change the event ID or some other part of the log record to make it look like it belongs. Another method is erasing the entries themselves. Some tools allow an attacker to erase certain log records that reveal their access attempts. The final kind of log cleaning tool we're going to look at performs this function in an obscure way. It populates the log with a large number of bogus entries to make searching for the attacker entries a long and drawn-out process.

Service Set Identifier

The Service Set Identifier (SSID) was never intended to be used to perform any type of security measures. The SSID's main purpose is for network identification as the name states. When a client's end device connects to the network, it has to have an identification setting to allow it to know what network to connect and operate on. When the wireless standards were created, the IEEE members had the fore-sight to realize that there may be more than one wireless network within range. This led to the creation of the SSID as a means to dif-ferentiate one wireless network from another. Today with the massive number of wireless networks, this has become a necessity.

The SSID can also be used to create multiple virtual wireless net-works. This is very similar to virtual LANs (VLANs) that are used in the wired world. Having multiple wireless networks does mean everyone is still sharing the same airspace, although they are on their own wired subnet. This is often used to accommodate guests and allow for different security levels. This can be seen in a situation where some older devices do not support the advanced security stan-dards. In this situation two wireless VLANs could be created, one with the advanced security open to go anywhere in the wired network and another that supports the older weaker security measures. The latter would also have some other wired security methods such as access contrtol list (ACL), IDS, IPS, or some other added security mechanism to balance out the higher wireless security risk it brings.

When we look at security with respect to the SSID, we see that most networks by default broadcast this information to anyone who is listening. As more and more people started to take a deeper look into the security of wireless, the thought of hiding the SSID in beacon frames was considered. It was noticed that if the SSID was not broad-casted, the existence of a wireless network can be somewhat masked. This masking would require the client to send a probe for an already configured wireless network. In all wireless networks, the existence of the SSID is easily attainable with some sort of wireless sniffer. This is because the SSID is part of the process of connecting to a wireless network and suppressed or not is still sent over the air during the connection process. This information can be read by any sniffing pro-gram, thus defeating any attempts to hide this identification informa-tion. Even with the SSID masked every time a client wants to connect to a network, they will send all of their connection settings including the SSID out into the air as part of their probing process. This can be seen in Figure 4.2, which depicts a wireless sniffer trace showing a non-broadcasted SSID.

Many vendors have default SSIDs that they program into their equipment. This is one of the first avenues that a hacker will take

Wireless SSID Sniffer Trace

802.11 Beacon	FC=........,SN= 648,FN= 0,BI=100,SSID=,DS=11
802.11 Probe Req	FC=........,SN= 589,FN= 0,SSID=ABC
802.11 Probe Req	FC=........,SN= 590,FN= 0,SSID=ABC
802.11 Beacon	FC=........,SN= 649,FN= 0,BI=100,SSID=,DS=11
802.11 Probe Req	FC=........,SN= 591,FN= 0,SSID=ABC
802.11 Probe Req	FC=........,SN= 592,FN= 0,SSID=ABC
802.11 Probe Req	FC=........,SN= 593,FN= 0,SSID=ABC
802.11 Probe Rsp	FC=........,SN= 651,FN= 0,BI=100,SSID=ABC,DS=11
802.11 Probe Rsp	FC=...N....,SN= 651,FN= 0,BI=100,SSID=ABC,DS=11
802.11 Probe Req	FC=........,SN= 594,FN= 0,SSID=ABC
802.11 Probe Rsp	FC=........,SN= 652,FN= 0,BI=100,SSID=ABC,DS=11
802.11 Probe Req	FC=........,SN= 595,FN= 0,SSID=ABC
802.11 Probe Rsp	FC=........,SN= 653,FN= 0,BI=100,SSID=ABC,DS=11

Figure 4.2   A wireless sniffer trace.

when trying to exploit a wireless network. Many companies also use very easy SSIDs such as Wireless, WLAN, and BRIDGE.

Shared Key Authentication

When connecting to a wireless network one must perform some type of authentication. Current IEEE standards provide two main types of authentication. The one we are going to look at first is shared key authentication. Shared key authentication was created to be the more secure of the two types; however, as we will shortly see, this actually became the less secure due to a small oversight in how it validates user keys.

Shared key authentication works via a challenge response mechanism. In order to explore this process we must first connect to the network. This is done by having the client device send out a probe frame. This frame will look for available wireless networks and their connection settings. Once an access point hears a probe it will respond with a probe response frame. This frame will identify all of its connection settings to the end device. In some cases, an end device will hear many responses from different access points in the area. To make sure that the end device connects only to the access point with the best signal, the probe response frame has a value for current signal strength. A client may hear multiple replies, although they will only connect to the access point with the highest signal strength value. Once the end client hears this and determines it supports the same settings as the access point authentication takes place.

The end device sends an authentication response frame to the access point. This frame is evaluated, and once the access point determines it is an authentication request, it will send a challenge packet back to the client. The challenge packet is made up of a clear text piece of data. The end device is required to encrypt this data with its wired equivalent privacy (WEP) key and send it back to the access point. Once this is done and the access point receives the packet it checks it against what it has for the encrypted version of that packet. If the results match, the access point will allow the end device onto the network. If the results do not match, the authentication fails and the end device is denied a connection.

Open Key Authentication

Open key authentication was originally seen as the less secure of the two authentication methods. The intent was to create an open network thus not requiring clients to have knowledge of the WEP key. As security became an increasingly visible issue, many vendors returned to the drawing board. Developing a solution that improved security while staying within the standard guidelines was difficult. These efforts led to the idea of using open authentication and unlike before this open authentication would require the use of a WEP key which was required to connect to the network. This worked because when you talked with the right WEP key your cyclic redundancy check (CRC) passed its test and the frame was allowed to access the network to its destination.

Looking at how open authentication worked, we see that the end device connected to the network as it did with a shared key. It makes a probe request, listens to probe responses from multiple access points in the area, and then determines the best access point to make a connection with based on signal strength. Open and shared authentication differ in the following ways. Open authentication sends an authentication request but does not receive a challenge; instead it is allowed to talk by default. When WEP is enabled, we have a slightly different process. When the wireless client starts to talk it automatically encrypts all the data with WEP encryption. When the access point hears data being sent it decrypts the frames and forwards them. If the frames were encrypted with a different key than the access point, the decryption portion fails and the packet is dropped.

Wired Equivalent Privacy Standard

The wired equivalent privacy (WEP) standard was created to give wireless networks similar safety and security to that of wired networks. WEP is defined as the optional cryptographic confidentiality mechanism used to provide data confidentiality that is subjectively equivalent to the confidentiality of a wired (LAN) medium that does not employ cryptographic techniques to enhance privacy. This gives us the basic thought of how WEP was created and what goals it originally intended to meet. In order to meet these goals, wireless had to address the three tenets of information security: confidentiality, availability, and integrity.

- The fundamental goal of WEP is to prevent eavesdropping, which is confidentiality.
- The second goal is to allow authorized access to a wireless network, which is availability.
- The third goal is to prevent tampering with any wireless communication, which is integrity.

To understand a WEP better we need to look at it more closely. The WEP protocol is used to encrypt data from a wireless client to an access point. This means the data will travel unencrypted inside the wired network. The WEP protocol is based off the RSA RC4 stream cipher. This cipher is applied to the body of each frame and the CRC. There are two levels of WEP commonly available, one based on a 40-bit encryption key and 24-bit initialization vector, which equals 64 bits, and one based on a 104-bit encryption key and 24-bit initialization vector, which equals 128 bits.

This protocol has been plagued with issues since its inception. A magnitude of exploits, poor design elements, and general key management problems have made WEP a very inadequate security mechanism. One of the original functions of WEP was to have the encryption unable to be affected by loss of the frame due to interference. What this means is when you send data across the air and lose the frame there would be no loss to the previous frame. With newer security methods and older wired secured methods it is common for subsequent packets to have an encryption dependency on the next or previous frame.

802.1x

The 802.1x standard was approved by both the IEEE and the American National Standards Institute (ANSI). On June 14, 2001, the IEEE approved the standard and four months later, on October 25, 2001, ANSI approved it as well. The 802.1x standard was designed for port based authentication for all IEEE 802 networks. This means it will work across Ethernet, fiber distributed data interface (FDDI), token ring, wireless, and many other 802 networking standards.

One thing people tend to get confused about is that 802.1x is in no way a type of encryption or cipher. All the encryption takes place outside the 802.1x standard. For example on a wireless network, the Extensible Authentication Protocol (EAP) would use one of its various methods of encryption for authentication. After the user is authenticated to the wireless network they may start a conversation using WEP, Temporal Key Integrity Protocol (TKIP), Advance Encryption Standard (AES), or one of the many other standard wireless encryption schemes. When we look at the 802.1x standard, at its most basic view we see actually what it was intended for, port based authentication. This means the standard takes your authentication request, decides if you are allowed onto the network, and then grants or revokes access.

Many parts of how 802.1x works is within other standards such as EAP and RADIUS. The 802.1x standard is just a mechanism that denies all traffic except EAP packets from accessing the network. Once the EAP protocol says it is ok for the device to access the network, the 802.1x protocol tells the switch or access point to allow user traffic. This is accomplished by having the network port, or in a wireless situation each client connection, in one of two port states: controlled and uncontrolled.

Figure 4.3 shows the three main designations in the 802.1x standard. Each of them has specific rules and functions. The standard was written to incorporate a large variety of different equipment; the names of these functions remain somewhat generic. The 802.1x protocol leverages two other standards. From the supplicant to the authenticator the standard is EAP. From the authenticator to the

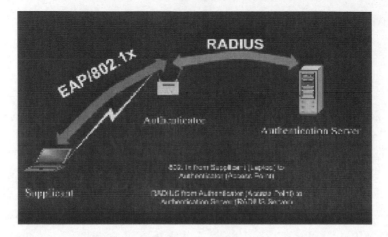

Figure 4.3   802.1x authentication process.

authentication server the protocol is RADIUS. The 802.1 x protocol takes EAP requests, sends them to a RADIUS server, and waits for an answer. Once this answer is received, it will allow or deny access to the network.

The authentication server, authenticator, and supplicant are the three main elements of any 802.1x exchange (shown in Figure 4.3). They each perform specific roles in processing the authentication exchange and allowing correctly authenticated devices or users onto the network.

Authentication Server

The authentication server provides the access granting and access rejecting features. It does this by receiving an access request from the authenticator. When the authentication server hears a request it will validate it and return a message granting or rejecting access back to the authenticator. This is the back end of the 802.1x standard and per the standard, the operation of this server is defined in another standard we will look at later in this section called RADIUS.

Authenticator

The authenticator is the first piece of network electronics that an 802.1x device will attempt to connect with. In our example, it is a wireless access point, although it can be anything providing access into the network. The device's role is to let only EAP packets pass through and to wait for an answer from the authentication server. Once the authentication server responds with an accept or reject message the authenticator acts appropriately. If the message returned is a reject message, it will continue to block traffic until the result is an access accept. When the accept response comes from the authentication server the authenticator then allows the supplicant the ability to access the network.

Supplicant

The supplicant is the device that wants to connect to the 802.1x network. This can be a computer, laptop, PDA, or any other device with a network interface card supporting the 802.1x standard. When the supplicant connects to the network, it has to go through the authenticator. This authenticator only allows the supplicant to pass EAP request traffic destined for the authentication server. This EAP traffic is the user's or device's authentication credentials. Once the authentication server determines that the user or device is allowed on the network it will send an access-granting message.

Extensive Authentication Protocol over Local Area Network (EAPOL)

Extensive authentication protocol over local area network (EAPOL) is a part of EAP, although it is outlined inside the 802.1x standard. Because of this, it is located in the 802.1x section in this book. This is because the 802.1x standard allows certain EAP message types to pass through an authenticator to the supplicant. The definition of messages that are allowed to pass through each message type and frame format had to be included inside the 802.1x standard. The EAPOL standard calls out the process and frame structure used to send traffic from the authenticator to the supplicant. This traffic is outlined with six frame types. This means only these six frame types are allowed to pass through an access point to a client. The IEEE created room for more, although the current standard only outlines six. Figure 4.4 shows each frame type along with the value that is used to identify the frame.

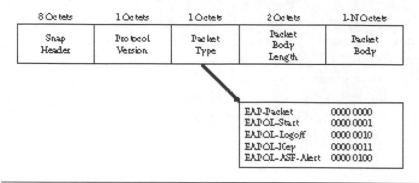

Figure 4.4 EAPOL frames.

EAPOL-Packet: This frame type is used to identify the packet as an EAP packet.

EAPOL-Start: This frame type is used to begin an EAP conversation or an 802.1x authentication.

EAPOL-Logoff: This EAPOL frame is used to end an EAP conversation or an 802.1x authentication.

EAPOL-Key: This EAPOL frame is one of the most security-related frames. It is used to exchange keying information between the authenticator and the supplicant.

EAPOL-Encapsulated-ASF-Alert: This is an EAPOL frame used to carry SNMP trap information out a non-802.1x authenticated port.

The most involved EAPOL type is the EAPOL-Key frame. This frame is used to send keying material like dynamic WEP keys. The only key frame defined in the 802.1x standard is the RC4 WEP key. As we get into 802.11i, we will see that some changes were made to the operation of the EAPOL-Key frame to accommodate other encryption cipher types outside of RC4.

Remote Authentication Dial-In User Service (RADIUS)

RADIUS stands for Remote Authentication Dial-In User Service; this protocol is used in network environments for authentication, authorization, and accounting. RADIUS can run across many types of devices such as routers, servers, switches, modems, VPN concentrators, or any other type of RADIUS-compliant device. The protocol works by creating an encrypted tunnel between the network device and RADIUS server. This tunnel is used for sending all the authentication, authorization, and accounting (AAA) information about who a user is, where they're allowed to go, and where they actually did go. In order to start this encrypted tunnel, a phrase or password called the shared secret is needed. The shared secret is located on the RADIUS participating network device and the RADIUS server. Once the shared secret is correctly set up secure communication can take place.

One of the benefits of RADIUS is the use of a common database of users to provide these AAA services across multiple device types. The database that RADIUS uses for storing usernames and passwords can be set up to point to many different types of directories. This means RADIUS can use most existing directory structures such as Microsoft Active Directory (MS-AD), Novell Network Directory System (NDS), Lightweight Directory Access Protocol (LDAP), and many other common directory types.

This protocol allows administrators to centrally locate and administer user access and accounting for all network equipment as well as remote access. RADIUS prevents many of the headaches associated with properly removing access to network equipment when employees are discharged. Once an organization has deployed RADIUS, user access can easily be removed in the event of a discharge. This was unlike the old days when administrators had to manually change usernames and passwords on all network equipment.

The RADIUS protocol specifications are currently defined in RFC 2865 and RFC 2866. RFC 2865 focuses on the access portion of RADIUS allowing user access into devices or onto the network. RFC 2866 focuses on the accounting portion of RADIUS allowing administrators to track changes and access to network devices as well as general access to the network itself. The original RFC numbers were 2138 and 2139; these were updated to address a number of security-related concerns. Another major reason for this update was to change the UDP port number of RADIUS from the original port numbers of 1645 and 1646 to 1812 and 1813. The protocol was changed because the UDP port number of 1645 was already designated by Internet assigned number authority (IANA) for the datametrics service and not for RADIUS. This is why most RADIUS servers support all four of these ports by default.

Now that we have covered the main usage and history of RADIUS, let us look at how RADIUS relates to wireless. RADIUS can be used as an access method to administer the access point. This is similar to how it would be used to administer routers or switches. The access point and the RADIUS server would have a shared secret and that would be used to set up an encrypted channel that can carry user authentication traffic. Another approach is to use RADIUS as explained in the 802.1x standard as a backend user authentication mechanism. RADIUS itself provides this feature so the 802.1x standard uses that instead of creating its own authentication mechanism. With this scenario, the access point would need to be set up correctly with the RADIUS server's shared secret and the access point would keep track of the user's request to enter the network. This means the user would only negotiate its authentication with the access point not the RADIUS server. This is similar to how RADIUS would be set up on a remote access device. A user would ask to enter the network; the user would then be prompted by the network device to provide some type of authentication. Once the user provides authentication the network equipment would verify it against the RADIUS server user database. If the credentials were correct, the user would be allowed onto the network; if it is not correct, the user would be denied access.

RADIUS has only four types of packets for authentication. Although there are other packet types for accounting, we are only going to focus on the authentication packets. The four types are as follows:

Access-Request: This packet allows the RADIUS sequence to take place.

Access-Accept: This packet informs the RADIUS client that the authentication provided to it was correct.

Access-Reject: This packet informs the RADIUS client that the authentication provided to it was incorrect.

Access-Challenge: This packet is used to challenge a RADIUS client for its authentication credentials.

Looking more closely at the RADIUS packet it should be noted that all four packet types are made up of the same packet format. They are identified by the code field. This field has a number of available codes that can be used; however, the only ones we are going to look at are the Access-Request identified by a one, the Access-Accept identified by a two, the Access-Reject identified by a three, and the Access-Challenge identified by a 12.

The next field, the Identifier, is used to match requests and responses to each other. This makes sure that multiple RADIUS conversations do not get mixed up as to what messages go to what device. The length field is used to identify the length of the packet. Because the RADIUS packet can have up two thousand attributes a mechanism to measure the packet length was needed. The last field holds the authenticator field, which is the field the password is protected in; this password is protected by a hashing mechanism.

RADIUS has a lot of complexity; according to RFC 2865 each RADIUS packet can be up to 4,096 bytes allowing 2,000 attributes in a single packet. When you set up RADIUS, it can be easy; however, depending on which vendor solution you select, there could be added complexity. Once thing to note is the security feature is not complex; time has shown if security is too complex it will be avoided or not installed.

When looking at the RADIUS server, as well as any authentication servers for that matter, the details around protecting the server itself is often forgotten. This stems from network people building and administering networks not servers. This leaves the server that RADIUS sits on exposed. For example if the network people build and administer networks, they most often do not have the skills to secure servers. You can have the most secure network in the world; however, if someone can easily hack into your authentication server your whole network is compromised.

Extensible Authentication Protocol

Extensible Authentication Protocol (EAP) is a standard method of performing authentication to gain access to a network. When Password Authentication Protocol (PAP) first came out, security issues quickly made it a less than desirable authentication method. After that Challenge Handshake Authentication Protocol (CHAP) came out and this also quickly became plagued with security issues. The industry decided it was easier to make an authentication protocol act the same way no matter how or what type of authentication validation took place. This meant for the first time a protocol could be inserted into products and software that allowed for passwords, tokens, or biometrics without having to write any extra code to support the different methods. This is how and why EAP was created. To use EAP you must specify inside the type field what kind of authentication you are going to use. This allows EAP to be used for passwords, tokens, and other authentication types. The EAP protocol can adapt to security issues and changes by leveraging different methods of authentication. EAP is also able to address new and always improving authentication techniques without having to make any changes to EAP supporting equipment.

When EAP was created a need for point to point protocol (PPP) compatibility was required. This helped ensure that the large variety of existing equipment could handle EAP without major modifications. In order to get this compatibility, EAP was included as a PPP type inside a PPP packet itself. This allowed for any device supporting PPP to be able to support EAP. EAP remained this way through RFC 2284. As EAP matured and required tighter integration with the 802.1x standard, its placement inside PPP was evaluated. The result of that evaluation is RFC 3748. Quoted below from RFC number 3748 is the reasoning behind the changes that were needed to EAP in order to support the 802.1x standard. "The IEEE 802 encapsulation of EAP does not involve PPP, and IEEE 802.1X does not include support for link or network layer negotiations. As a result, within IEEE 802.1X it is not possible to negotiate non-EAP authentication mechanisms, such as PAP or CHAP [RFC1994]."

One of the main points of using EAP is the ability to leverage multiple types of authentication mechanisms. This has helped to prevent EAP from becoming obsolete due to security vulnerabilities or protocol weaknesses. The ability to use multiple authentication types is located in the type field on an EAP packet. The original standard as well as the new RFC 3748 only listed three main EAP types: MD5 Challenge, One Time Password (OTP), and Generic Token Card (GTC). Today there are a number of different EAP types, some of which are vendor specific, some detailed in the EAP standard, and some detailed within their own standard document. Others are industry standards on their own detailed by Internet engineering task force

(IETF) documents or RFC documents. In the section below, the most widely used, wireless-related EAP types will detailed and examined.

EAP-MD5

The EAP-MD5 (Extensible Authentication Protocol—Message Digest version 5) is one of the most limited EAP types included in the EAP RFC. This version uses a message digest (MD) hashing algorithm to validate user credentials. Some of the other types of EAP methods create encrypted tunnels and then inside these tunnels they perform EAP-MD5 validation. One of the requirements of EAP-MD5 is a shared secret. This secret needs to be shared out of band so two parties share the secret. That secret is then used to encrypt a challenge to verify that the other party has the same secret.

EAP-TLS

The EAP-TLS (Extensible Authentication Protocol—Transport Layer Security) method is described in RFC 2716. It was created by Microsoft in October 1999. The RFC was built off RFC 2284 for PPP and RFC 2246 for TLS. TLS came about from the older secure socket layer (SSL) protocol. Netscape created SSL and used it for secure web browsing. Once the Internet became popular, updates to SSL were required. In 1996, the IEEE created TLS based off Netscape's SSL and Microsoft's private communications technology (PCT). This EAP method uses certificates to authenticate users and requires certificates at both the server and client end. This is where TLS plays into this standard because it already provided a good way to perform the needed certificate management steps. This particular EAP method is one of the strongest, although it prevents usage unless you are accessing the network from a computer with your client certificate already installed on it.

Setting up a wireless network with 802.1x and EAP-TLS requires some upfront work and planning. First, you must have a certificate authority (CA); this server will function as the distributor of both client and server certificates. Also needed would be an AAA server that supports EAP-TLS type. Finally, you need a client that can support this EAP type. Once all the parts are in place, the next challenge of correctly configuring each part to interact is necessary.

EAP-TTLS

The EAP-TTLS (Extensible Authentication Protocol—Tunnel Transport Layer Security) is an IETF draft document created by Funk Software Inc. In 2005 Funk Software Inc. was acquired by Juniper Networks, Inc. The latest version to date is named draft-ietf-pppext-eap-ttls-05.txt and was created in July 2004. The reasoning behind creating a new EAP type was based on an opportunity Funk Software saw in the market. This opportunity was a need to support older devices that were not able to perform the new authentication types. This gave Funk the idea to write an EAP type that allowed for secure communication of credentials along with the ability to allow legacy authentication types.

LEAP

LEAP (Lightweight Extensible Authentication Protocol) is a Cisco Systems' proprietary protocol. Cisco did release the source code for vendors who wanted to incorporate LEAP into their wireless adapters. The list of vendors includes D-Link, Dell, SMC, 3Com, and Apple. The code for LEAP is still considered Cisco Systems' intellectual property and is available for use only under a non-disclosure agreement (NDA).

PEAP

PEAP (Protected Extensible Authentication Protocol) was created as a joint effort between RAS, Microsoft, and Cisco Systems. Currently PEAP is in an IETF draft called draft-josefsson-pppext-eap-tls-eap-08.txt last updated in July 2004. Because it still lives in draft form updates may change its version number or document name. PEAP was a move by the industry to make a single EAP method that multiple vendors could share. The three vendors who created the standard implemented it each in their own way; this made Microsoft, and Cisco versions of PEAP different and not interoperable. This has been slowly working itself out of the system, although do not expect an easy integration when trying to use PEAP methods from Microsoft and Cisco interchangeably.

One of the main advantages of PEAP is the ability to have a strong EAP type that does not require client certificates like EAP-TLS. PEAP works similarly to EAP-TLS by creating an encrypted tunnel with TLS and then performing another EAP method inside this encrypted tunnel. Unlike EAP-TLS when PEAP performs this process, it does not validate a client certificate. This is where Cisco and Microsoft differ; each of them uses a different method after the TLS connection is created.

EAP-FAST

EAP-FAST (Extensible Authentication Protocol—Flexible Authentication via Secure Tunneling) is an IETF document created by Cisco Systems in February 2004. The current document is named draft-cam-winget-eap-fast-00.txt and is located on the IETF website. Cisco had some security flaws released on their LEAP method and instead of fixing them they abandoned the proprietary standard and created this EAP method. This EAP method supports a fast roaming time compared to the other EAP standards. This timing was a critical requirement for many companies that have Wi-Fi phones or other time-sensitive application. This made Cisco Systems in need of a fast secure EAP method for wireless authentication.

Wi-Fi Protected Access

The Wi-Fi Protected Access (WPA) standard has an interesting history in relation to how it came to be a standard. When the security of WEP was broken, the industry turned to the IEEE to fix it. The IEEE said it would create the 802.11i wireless security standard but developing this standard took a long time. As it took longer and longer to be ratified, sales of wireless devices declined because of a lack of a standard secure wireless networking method. Wireless manufacturers started to push the IEEE and other standards boards to ratify the standard so they could produce secure products. With the delay in the 802.11i release date the Wi-Fi Alliance decided that they would create a subset 802.11i standard called WPA. The Wi-Fi Alliance created WPA by leveraging what the 802.11i task group had already done and formalizing it into WPA. This meant that any large changes to the 802.11i standard would influence future versions of WPA. This was seen with WPA and WPA2. Today with 802.11i complete, the use of WPA has been greatly reduced.

The WPA standard supports two methods of authentication and key management. The first one is EAP authentication with the 802.1x standard. This method works through the use of the 802.1x protocol and a backend authentication server. It leverages EAP for over-air authentication and RADIUS for backend authentication. This method is the most secure of the two and provides the lowest amount of end client administration.

The next available option is to use preshared keys. This option requires a key to be applied to the devices and the wireless access points. This also means that everything has the same password entered into them. To combat someone using this key to eavesdrop on other conversations WPA uses a method that creates a unique session key for each device. This is done by having a preshared key called the group master key (GMK) that drives a pair transient key (PTK). How this works will be explained in the section on 802.11i. This second solution was added to WPA for home and small office support. In a house or small office, you are unlikely to have an authentication server such as RADIUS. A PSK is a 256-bit number or a passphrase 8 to 63 bytes long. WPA does support TKIP and Message Integrity Check (MIC) for older devices.

One of the reasons why 802.11i was not ratified was because of certain requirements that were not well defined at the time. With the WPA standard using whatever the IEEE 802.11i task force had already completed, some changes were needed to be able create this interim standard. These changes led to a number of differences between the two standards.

The first big difference is WPA supports TKIP by default. This is unlike 802.11i, which supports AES CCMP by default. The next item is the fact that WPA does not even support AES CCMP. The last major item that differentiates WPA and 802.11i is the RSN IE. This is used to pass the supported cipher settings between wireless access points and clients. In 802.11i, this portion was not well defined so the WPA standard had to create some rules without having them affect anything that might be done to the RSN IE from the 802.11i task group. This was accomplished by creating a WPA IE and using different values to distinguish them from one another. This helped so that once the RSN IE was well defined it was not hard to put into WPA.

802.11i

In this section, we are going to look at the ratified 802.11i security standard. This standard came about from a need to improve the security of 802.11 networks to a level sufficient to warrant wireless as a generally accepted secure transport media. In this standard, the IEEE outlined a secure way to access wireless networks. They also tried to mitigate the enormous number of threats that were making wireless networking a real risk for companies.

Looking at the process of how 802.11i became a standard, we need to go back to July 1999 when there was interest in enhancing the MAC layer of 802.11 for quality of services (QoS) and privacy. This built up enough steam to create a task group TGe in March of 2000. After a year it was determined that this group needed to be split into a security and a QoS group because of the large workload. This spilt created the TGi security working group. This group created the 802.11i standard and put it up for approval by voting on it. In order to proceed to the next level of voting the working group had to have a 75 percent approval vote. This approval took a number of drafts over a three-year period. Once this was done, there was a level of sponsor ballots and finally the standard boards' approval process. On June 24, 2004, the standard body finally approved the 802.11i standard.

802.11i uses a number of standards, protocols, and ciphers, which have already been defined outside of the 802.11i. A number of standards are also defined inside it as well.

RADIUS, 802.1x, EAP, AES, RSN, TKIP, and others are some of the defined standards that are part of the 802.11i standard. Some of these are defined inside their own document and some of them are officially created inside the 802.11i document. The first portion of 802.11i that we are going to talk about is the Robust Security

Network (RSN) standard. This standard is used for dynamic negotiation of authentication and encryption. This is used to negotiate what kind of encryption a client can support as well as what type of encryption is required based on a policy.

Another piece of the 802.11i standard is the ability to use EAP. It was determined that the 802.11i standard would not specify an authentication method or type; rather it would allow a protocol that can perform multiple types of authentication inside itself. This is exactly what EAP does, it allows many different authentication types from passwords, smart cards, certificates, and many others to be used based on the same request, accept, and reject methods. For EAP to work correctly with the 802.11i standard another well-known standard had to facilitate the transmission of EAP between untrusted and trusted entities. This is where the 802.1x standard fit. Its main goal is to provide a framework for strong authentication and key management. The 802.1x protocol allows the access point only to allow an EAP request into the network. This is the case until the client is properly authenticated. Once this is done, key negotiation and subsequently network access can be achieved.

As WPA was included, the 802.11i standard needed to have an option for environments where an authentication server was not financially acceptable. This authentication server was a requirement of the 802.1x standard. In order to make 802.11i viable for both large enterprises and small office home office (SOHO) users another method needed to be created. This is where the pre shared key method came from. This is very similar to WPA and its preshared key method. When a preshared key is used, each client uses a secret to create subsequence-keying material. This master key is the same across the network like WEP, although it is used to create a session-based key for each client.

Before we get into the details of how 802.11i works, we need to understand all the components making up 802.11i so we are comfortable with a full-system overview. Below is each main portion of the 802.11i standard. As said above, most of the standards and protocols inside 802.11i are either their own standards or are located inside the 802.11i standard. The sections below outline the standards that are located inside the 802.11i standard.

Robust Secure Network (RSN)

As part of the 802.11i security standard, robust secure network (RSN) was created. RSN specifies user authentication through IEEE 802.1X and data encryption through TKIP or Counter Mode with CBC-MAC Protocol (CCMP). RSN also has the option to use TSN in order to use older security methods such as WEP. TSN will be explained in detail in the next section. RSN uses the TKIP and AES for encryption to protect the confidentiality of data. The TKIP solution is used for a backwards compatibility for legacy devices and the AES is what RSN is using as a long-term encryption method. The way AES is set up is in Counter Mode with CBC-MAC Protocol (CCMP). AES can be set up and used in multiple ways so the 802.11i standard states that AES must be used in a method called CCMP. The RSN protocol also uses EAPOL-Key messages for key management. In this section we're are going to see how RSN works with 802.11i in aiding to choose an available authentication method and encryption cipher scheme.

Advertising the cipher suites supported on an access point and client is done through robust secure network association (RNSA) messages. These messages spell out the supported ciphers of each party and negotiate what method will be used to connect securely. These messages are located inside what is called an RSN IE or RSN information element. An RSN IE is used to tell the other devices about what cipher suites the sending device supports. The RSN IE can be sent in a beacon from an access point or in an association request from a client. After an association request, a response will be returned with an RSN IE listing what requesting method matched the method supported by the other party.

The standard allows RSN IE optionally to be inside each of the following management frame types:

- Beacon
- Association Request
- Reassociation Request
- Probe Response

Of the eleven sections of the frame only the first three are required in all RSN IE transmissions. After the three required fields are present, all other fields must have the preceding field inserted with data or the frame will not be properly received. This means that if you need the value in the ninth field you have to include all the other preceding eight fields.

Element ID: This field supports 48 decimals or 30 hexadecimal digits. Currently the only allocated element ID is 48, which stands for RSN.

Length: The section identifies the total length of the RSN IE frame. Currently the frame is only 255 octets long. When using a large number of cipher suites you may run into a case where you can only support a limited number of cipher suites. This is due to the limit in the total size of the RSN IE frame.

Version: The version field is used to show what version of RSN is currently being used. It holds up to two octets. Today only, a single version of RSN exists. This version is shown as one in this field. Version 0 and 2 are reserved for other versions.

Group Cipher Suite: This is the cipher suite used to protect broadcast and multicast traffic. It holds up to four octets of information about the total group cipher suites used. This field holds up to two octets of information about the multicast and two octets for broadcast cipher suites used.

Pairwise Cipher Suite Count: This field lists the number of selected pairwise cipher suites. This field is only two octets long.

Pairwise Cipher Suite List: This field contains all the ciphers that were selected for the pairwise key. Each of the cipher suites are accounted for in the pairwise cipher suite count field. Each one has a corresponding type inside this field. Each one of the used cipher suites is four octets long. Of the four octets, three are used for the organizationally unique identifier (OUI) field and a single octet is left to identify the cipher suite.

AKM Suite Count: The authentication and key management suite count is used to determine how many different key management options are available, such as preshared keys or ones dynamically allocated with 802.1x. In an IBSS only a single AKM suite can exist. This field has a maximum size of two octets.

AKM Suite List: The authentication and key management suite list is used to specify what key management options are available. Depending on the AKM count suite field, this field could have multiple four-octet sections defining each key management option. Currently only two options exist, pre-shared keys and 802.1X key management. This leaves a number of reserved and vendor-specific options to include later.

RSN Capabilities: This is a two-octet field used to identify what RSN capabilities are available on the network. It identifies if the device is capable of a pairwise key. It also allows the device receiving the RSN IE to understand if it can support RSN; if it cannot support RSN TSN can be tried. In the event that it is not understood, it is assumed that it cannot support RSN.

PMKID Count: The pairwise master key identifier count field is a two-octet field used only with reassociation. It is used to cache keys so when a client is roaming it does not have to go through the entire authentication process with each access point. This is to speed up the timing and lower the bandwidth as a client roams from one access point to another. The count is to define how many of these credentials are currently inside the RSN IE frame.

PMKID List: The pairwise master key identifier list field is four octets for every PMKID identified with the PMKID Count Field. This is where each of the different types of PMKID is stored. Currently there are three main PMKID types listed. The first one is a cached PMK that has been obtained through pre-authentication with another AP. The second one is a cached PMK from an EAP authentication. Last is a cached PMK from a PSK.

In RSN, we are only concerned about telling the other party what cipher suites are supported. Once that has been answered, we can decide if it is possible to negotiate a common cipher suite or method between both parties. Details about each cipher suite we're about to look at either have already been explained in detail in a previous chapter or will be explained in upcoming sections of this chapter. In the RSN cipher suite frame section, we have six specified cipher suites and a number of reserved and vendor-specific cipher suites that can be used in future. Today the six supported cipher suites are identified with the following hexadecimal code 00:0F:AC.

The RSN standard is a method to negotiate what type of security methods are supported by each client and each access point. These security methods are identified as cipher suites inside of the RSN IE frame. These cipher suites allow for the use or nonuse of any combination of security methods. This means a policy could be put into place that negates the use of weaker security methods such as WEP and allows for a choice of TKIP or AES. This gives the architect or designer the ability to create a policy allowing or denying whatever cipher suites he or she might feel are weak or not needed.

Transition Secure Network (TSN)

The transition secure network is part of the RSN portion of 802.11i. It is used to achieve backwards compatibility with older wireless systems. It was broken out from RSN to provide this backwards capability. With RSN it is possible to have a number of authentication and encryption types running on an access point. To make sure that some of the weaker authentication and encryption types were not set up in RSN they were taken out and considered TSN. This makes RSN more secure and allows an easy way to turn off all the older weak methods. With RSN if you select not to support TSN, WEP will not be included as an option to negotiate between the access point and wireless client.

Temporal Key Integrity Protocol

The Temporal Key Integrity Protocol (TKIP) was an interim solution developed to fix the key reuse problem of WEP. It later became part of the 802.11i and subsequently part of the WPA standard. This meant there were various flavors of TKIP until 802.11i was finalized.

TKIP was included in the 802.11i standard for backwards compatibility. The 802.11i standard did not want to use a cipher based off RC4 so they chose AES. We will talk about AES shortly until then just realize that TKIP was put into 802.11i for the sole reason to help older devices transition to 802.11i. To do this 802.11i needed to support a protocol that could easily upgrade WEP to something safe enough to include in 802.11i. One of the main reasons for using TKIP over WEP came from the increased security and increasing number of attacks that were plaguing the WEP protocol. Using TKIP protected against these attacks and reduced the overall risk of operating a wireless network.

The industry also saw value in the TKIP standard because the migration from WEP to TKIP was an easy one. In most cases, moving from WEP to TKIP was a small firmware change. This meant that no hardware was required to make the change and it meant most older, already purchased equipment would be able to upgrade to TKIP.

Another interesting note about TKIP comes from Cisco Systems. Cisco came up with a TKIP solution well before the 802.11i standard defined one. This led some people to wonder about which versions of TKIP is on certain products. Other vendors outside of Cisco also created TKIP-based solutions before the standard was ratified. Today Cisco differentiates their versions of TKIP with the standard one by calling theirs Cisco Key Integrity Protocol (CKIP). In Cisco products, you can specify to use TKIP, which is the 802.11i compliant version, or CKIP, which is the Cisco-created version.

The TKIP encryption portion works in a two-phase process. The first phase generates a session key from a temporal key, TKIP sequence counter (TSC), and the transmitter's MAC address. The temporal key is made up of a 128-bit value that is similar to the base WEP key value. The TKIP sequence counter (TSC) is made up of the source address (SA), destination address (DA), priority, and the payload or data. Once this phase is completed a value called the TKIP-mixed transmit address and key (TTAK) is created. This value is used as a session-based WEP key in phase two.

In phase two, the TTAK and the IV are used to produce a key that encrypts the data. This is similar to how WEP is processed; the first 24 bits of the IV are used in the WEP and are sent in the packet header. TKIP extended the IV space allowing for an extended IV field, which holds an additional 24 bits. In phase two, the first 24 bits are filled with the first 24 bits of the TTAK. The next 24 bits are filled with the unused portion of the TSC. This is safer than WEP because the key is using a different value depending on who you are talking to. In WEP, the same random value is created by each client or access point. Some products never even created a random value and just incremented the value by one making it an easy target for hackers.

The bases of TKIP came from the WEP protocol. In the 802.11i standard, TKIP is referred to as a cipher suite enhancing the WEP protocol on pre-RSNA Hardware. This is said because RC4 is still used as a cipher although the technique in which it is used was improved greatly.

TKIP MIC

Next, we are going to look at the Message Integrity Check (MIC). Similar to TKIP there were also many versions of MIC before 802.11i defined it as a single standard. Once this was done, MIC became known as Michael although the name MIC remained in use. Today with 802.11i ratified MIC is Michael and vice versa. The protocol itself was created to helped fight against the many message modification attacks that were prevalent in the WEP protocol. The IEEE 802.11i standard describes the need for MIC in the following quote "Flaws in the IEEE 802.11 WEP design cause it to fail to meet its goal of protecting data traffic content from casual eavesdroppers. Among the most significant WEP flaws is the lack of a mechanism to defeat message forgeries and other active attacks. To defend against active attacks, TKIP includes a MIC, named Michael." The MIC was created as a more secure method of handling integrity checking compared to the IVC in WEP.

The MIC is a hash that is calculated on a per packet basis. This means a single MIC hash could span multiple frames and handle fragmentation. The MIC is also on a per-sender, per-receiver basis. This means that any given conversation has a MIC following from sender A to receiver B and a separate MIC flowing from sender B to receiver A.

The MIC is based on seed value, destination MAC, source MAC, priority, and payload. Unlike IC, MIC uses a hashing algorithm to stamp the packet giving an attacker a much smaller chance to modify a packet and have it still pass the MIC. The seed value is similar to the WEP protocol's IV. TKIP and MIC use the same IV space, although they have added an additional four octets to it. This was done to make the threat of using the same IV twice in a short time less likely.

The MIC is also encrypted inside the data portion, which means it is not obtainable through a hacker's wireless sniffer. To add to this the TKIP protocol also left the WEP IVC process which then adds a second less secure method of integrity checking on the entire frame. To combat message modification attacks the TKIP and MIC went a step further and introduced the TKIP countermeasures procedures. This is a mechanism designed to protect against modification attacks. It works by having an access point shut down its communications if two MIC failures occur in sixty seconds. In this event, the access point would shut down for sixty seconds. When it came back up it would require that all clients trying to reconnect change their keys and undergo a re-keying. Some vendors allow you to define these thresholds, although the TKIP MIC standard calls out these values.

To prevent noise from triggering a TKIP countermeasure procedure the MIC validation process is performed after a number of other validations. The validations that are performed before the MIC countermeasure validation are the frame check SUM (FCS), integrity check some (ICV), and TKIP sequence counter (TSC). If noise was to interfere with the packet and modify it one of these other checks would be able to find it first preventing the frame from incrementing the MIC countermeasure counter.

Advance Encryption Standard

Advance Encryption Standard (AES) can be applied in many different ways. The way that the 802.11i standard has chosen AES is with CCMP, which is based on CBC-Message Authentication Code. It was chosen for data integrity and authentication with the Message Authentication Code (MAC) providing the same functionality as Message Integrity Check (MIC) used for TKIP. Before we can get into CCMP, we need to look at AES and some of its modes. The first term is CTR; this is AES in counter mode. This mode is used for confidentiality. The next mode is called CBC-MAC, which stands for cipher block chaining message authentication mode. This mode is used for integrity. AES also has combined CTR and CBC-MAC to create CCM. CCM stands for CTR/CBC-MAC mode of AES that incorporates both the confidentially of CTR and the integrity of CBC-MAC.

802.11i System Overview

Now that we have looked at each part that makes up the 802.11i standard, we can look at the standard as a whole. As we go over this, we will see how the client connects to the access point, authenticates, and negotiates keys. Each one of these steps leverages the outlined standards we have to talk about.

The client would first need to make a connection to the access point. This would happen though the normal open key authentication process we saw above. Contrary to most 802.11 standards 802.11i only allows for open system authentication. This is due to a security flaw in shared-key authentication that was discovered.

After the initial connection request, the client would need to hear a RSN IE broadcast or send a probe request with an RSN IE. Whichever way this RSN IE frame is sent, both clients will need to negotiate on a cipher suite for use. After the RSN IE frames are sent and a negotiation is reached the EAP process starts. This can start with the access point sending an EAP identity request or a client sending an EAPOL-Start frame. Once the EAP process has started, it will go through the EAP authentication process that is associated with each particular EAP type. It ends with the client receiving an EAP success message from the access point. During this process, an AAA key is sent from the authentication server to the wireless end device. This key is used as a seed key to create the keys outlined below.

The key exchange process takes the original 802.1x EAPOL-Key frame and makes some modifications allowing for the use of WEP-40, WEP-104, TKIP, and CCMP cipher suites. From the 802.1 x section, we saw that the EAPOL-Key frame only supported WEP-40 and WEP-104 keys. The 802.11i standard modified this and added the ability for the frame to carry TKIP and CCMP keys as well. This key exchange is accomplished by a process known as the four-way handshake. This process takes two main keys and creates unique group and session keys for each client. These session and group keys are created from the two main keys. The main keys are known as the pairwise key or the pairwise master key (PMK) and the group key or the group master key (GMK).

In an 802.1x 802.11i setup, the PMK comes from the authentication server. If the 802.11i setup is using preshared keys then the

PMK is mapped to a password. The PMK is divided into three keys. The first key is the EAPOL-key confirmation key (KCK), which is used to provide data origin authenticity. The second key that is created from the PMK is the EAPOL-key encryption key (KEK), which is used to provide confidentially. The last key is the PTK, which is also used for data confidentiality. To create the PTK a pseudorandom function takes place with the access point's MAC address, client MAC address, and a nonce sent from each side as well. This allows a single master key to create multiple session keys without having to re-exchange a new master key each time.

The next key is the group key or group master key (GMK). This key is similar to the PMK except that it is used for beacon and management traffic encryption. The same process of hashing senders and receivers MAC addresses and nonce's are used to create a group temporal key (GTK) from a group master key.

Now that we have talked about the keys and how they are split up to accommodate session encryption, we need to look more closely at the four-way handshake we discussed. This handshake starts with the authenticator sending the supplicant a nonce. This is often referred to as the ANonce in the 802.11i standard. This nonce is a random value used to prevent replay attacks. This means that old nonces cannot be reused. After each party receives a message, the first step before any other is a check to see if the nonce was changed or if the same nonce was incorrectly reused. Once the first message is received by the wireless client, it will check the nonce and then generate a SNonce. This nonce will be used in the next step of calculating the pair transient key (PTK). After the PTK is created, the client will then send the SNonce as well as the security parameters outlined in the RSN IE frame to the access point. This information is the second message in the four-way handshake. All of this information will be encrypted using the KCK, which will protect it from any modification while in transit. Once this is received by the access point, it will check that the nonce is not an old value. Once this is done, it will also generate the PTK from the SNonce and ANonce and then check the KCK to make sure it was not modified in transit. Once this is done, the third message in the four-way handshake will take place. This message is used to tell the client to install the PTK key that was created and if used this message will send a GTK to the client to install. Once the client receives this it will check

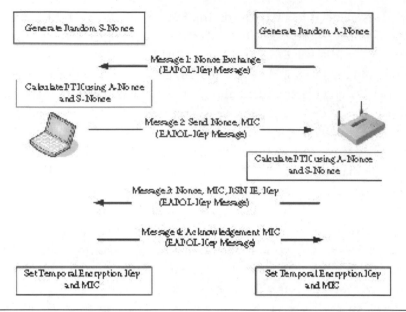

Figure 4.5   Four way hand shake.

the KCK and if it is correct install the key or keys. The last message is a confirmation used to let the authenticator know that the client has successfully installed the keys and is ready to communicate using them. Figure 4.5 shows this four-way handshake as it takes place.

Wi-Fi Protected Access

After 802.11i came out the Wi-Fi Alliance wanted to continue the initial investment made in WPA. This posed an issue since the 802.11i standards were now out and another standard was not what the industry needed. In order to keep WPA going they decided that they would go back to the core benefit their organization provided, standard interoperability testing and certification. In creating WPA2 the Wi-Fi Alliance made this version of WPA an interoperability mark similar to Wi-Fi. This mark ensures that any product carrying it has an interoperable 802.11i standard.

Rogue Access Points Detection

Detecting rogue access points has become almost an industry of its own. There are a number of products that perform rogue detection, from access points with built-in detection capabilities to appliances to stand-alone devices whose sole purpose is rogue detection.

Each company that developed rogue access point detection software took a unique approach in doing so. This has given us a wide variety of techniques we can use to detect and combat rogue access points.

Some of the products have the ability to detect and prevent rogue access points from connecting to the network. How this works is by using other access points that are managed by the same product. These access points listen for beacons or data frames coming from other access points that are not defined in the console. When this happens, the unknown access points are labeled as rogue. The next step leverages the wired network to shut down rogue access points. This is done by querying the switches in an attempt to find the rogue access point's MAC address. When the query is made to the network the switch, which has the access point's MAC address in its CAM table, it will respond, saying I have that MAC address. The next action can be to tell the switch to shut down that port and thus disconnect the access point's connection to the network. This method can fight a moving rogue access point very well. As soon as someone moves the access point to somewhere else the whole process starts over, finding, identifying, and then shutting down the rogue access point. Many newer access points make this method a little difficult as they have various MAC addresses for both the wired side and the wireless side.

Some of the other approaches that have been made to go beyond the location and identification of rogue access points have to do with using denial of service attacks. This feature is interesting and maybe considered illegal. This is definitely something that someone in a multiple tenant location may want to consider before using. This approach works on the basis of finding a rogue access point and then using nearby access points to send disassociation messages to all clients connecting to the rogue access point. As we have said in a multiple tenant location, this could mean disrupting a neighbor's network. This could become a legal issue quickly.

Wireless Security Tools

Wireless networks pose a threat to all who use them. Wireless networks lack the safety of having your communications securely transmitted inside physical cabling. With wireless communications, the air is your transmission medium and that medium is accessible to anyone who wants to listen. This has led to the development of many tools to aid the curious in eavesdropping on other people's wireless communications. These tools are most often created for network management or troubleshooting and include some powerful capabilities. Often these same tools have a very dark side if used for that particular purpose. Some of the tools were made for the sole purpose of attacking wireless networks. People often ask why these tools are made and why legislation has been put into place to stop them. The answer to this question is one that is often debated. The most common answer to this question is that these tools educate and make the public aware of what an attacker can do and how easily they can do it. Most hacking tools stem from a vulnerability or flaw that already exists. All that the tool writers do is make it easy for people to perform the same type of exploit without having the knowledge level required to find it in the first place. Face it, if you can download a program, run it successfully, and defeat your security you might want to start thinking twice about who else might want to try the same steps. Many people have debated the value of releasing tools that circumvent or break security methods. No matter what they say almost all tools come from an already existing security problem not a new one created by the tool.

The tools used on wireless networks have a few main functions made up of scanning, sniffing, cracking, and causing a denial of service attack. Numerous tools out today do the exact same functions as other available tools. In this section we will address the main functions of the various tools out on the market or in the wild.

Scanning Tools

Scanning tools are easy to use sniffers that capture network data and provide a GUI for the user to view information relating to wireless network identification. The scanners make it easy for someone to see the type and settings of a wireless network. This is very useful for someone looking for networks in a war drive. The main function of what a wireless scanner does is to find wireless identification information. Some of the information that these scanners will find are channel, security type, MAC address, access point vendor, data rate, signal strength, noise level, signal-to-noise ratio, and GPS location. Other than GPS all of the above information can be found with a wireless sniffer. Scanning tools allow the user to see what networks are in their area and what their settings are very easily. Even though sniffers can find this information, these scanners do a much better job of presenting the relevant information to the user. These kinds of wireless tools are what are commonly used for wireless network identification. Most war drivers use these wireless scanners to find networks and then move to a sniffer to get information once they have selected a target.

Sniffing Tools

Sniffing tools allow the user to view packet data on a wireless network. This gives users the ability to see network traffic. This traffic can be seen in real time based on the type of sniffing tool used. There are many wireless sniffers on the market today. Some range from high-end enterprise type sniffers that cost thousands of dollars to free open source sniffers.

Sniffers are a great tool and can help tremendously in troubleshooting networks, although in the wrong hands a sniffer can be used for many malicious purposes. If an attacker uses a sniffer, they can see lots of information about your wireless network. They can see the presence of a wireless network and its SSID, channel, power, and data rate information. With this information, an attacker can connect to an insecure network or understand what type of encryption a network may have. In addition, the sniffer can find IP-related information, which can help an attacker with network mapping functions. This can be expounded by evaluating high traffic talkers based on source and destination. This information can help an attacker find a file server or a critical application based on the amount of traffic sent to it. Most commonly, an attacker will use a packet sniffer to start to mount a man in the middle attack to circumvent any existing security measures. Once this is done, they can get to the next step of performing an exploit and seeing the data unencrypted.

Network sniffers can also find other interesting information relating to weak security methods of older widely used network protocols. A number of existing protocols have clear text authentication set as default. Telnet is a good example of one; if an attack had a sniffer set up, they could filter for this type of traffic and just capture it. When a network administrator tries to connect to a piece of networking equipment over the wireless, they will be prompted to enter their username and password. Once the authentication takes place, the sniffer will see the clear text authentication and record the username and password. Email is another one of the other problem cases relating to clear text authentication. Pop3 email retrieval is authenticated in clear text as well. This becomes an issue for mobile workers pulling email through a wireless network in a coffee shop or other hotspot. When they request their pop3 email, they have to authenticate in clear text to receive it.

Hybrid Tools

When we defined wireless tools into scanners, sniffers, and crackers, we still needed some way to classify the tools that performed multiple functions. This is why there is a hybrid section. Tools that can easily present a GUI with relevant wireless detection information are called scanners. If they have in-depth packet analysis features, they are sniffers. If they can display an easily readable GUI and have in-depth packet analysis under the hood they are hybrid tools. Some of the most valuable tools in any wireless professional's tool kit are listed in the hybrid section.

Cracking Tools

The wireless cracking tools we are going to talk about are used to break network encryption types on wireless networks. They work by taking advantage of the encryption type or the method in which the encryption is applied. Most ciphers have strong cryptographic functions, although the mechanisms used to implement the cipher often have flaws that are identified and then exploited. This section will identify some of the widely used cracking tools for wireless networks.

Access Point Attacking Tools

This section will detail the tools that can be used to attack the access point itself. These tools are used to exploit many widely used Internet protocols for remote management and monitoring. These tools are not directly related to wireless, although they will still defeat the security of an access point. Most of these tools take advantage of problems that exist in widely used protocols such as SNMP, HTTP, and TELNET.

Wireless Security Policy Areas

A wireless security policy should encompass all wireless technologies inside an organization. With the lines rapidly blurring between phones, laptops, and PDA devices a policy covering all wireless communications will help prolong its lifecycle and adapt to newer technologies more easily. Every company, including ones not using wireless networking technologies, should have a wireless security policy. Having a wireless security policy in a company not using wireless networking technologies enforces the fact that certain wireless technologies are not allowed. This sets the expectation of what will happen to employees if wireless communication technologies or methods not approved in the policy are being used.

Policies are hard to write for companies who already have in-depth policies. In this environment, many of the basic policy points are addressed in their own documents. For example, most companies have a password policy. In a wireless security policy, passwords are often used. In order to keep everything from contradicting, two actions can be undertaken. First is to update all general policies that wireless can affect; this means updating the password policy, acceptable use and abuse policies, and many others. This makes the wireless security policy scattered across many other policies making it hard to update and maintain.

The second option is to create subsections in the wireless security policy. These subsections will address each of the areas where there are other policies already in place. These other policies can be referenced or left out completely. As long as no contradictions take place, keeping these subsections inside the wireless policy will help in the maintenance of this policy. On the other side of this approach is the level of effort involved in ensuring that if any of the other policies are updated that the wireless policy does not create any contradictions. Each company will need to make their own decisions about where and how to address each area of the wireless security policy.

The first step in policy development should always be a risk assessment. Once the risk assessment is complete and upper management has relayed their top priorities for the policy, the writing portion begins. The next few sections address the main areas that should be included in a wireless security policy. Before we get into each area, a level set has to take place in the policy stating what it is, what it is for, and whom it affects. In some large organizations, policies are often not company-wide. When this is true an explanation of the area that they enforce is needed.

Password Policy

The use of passwords has always been part of information security. We have seen many uses for passwords and many ideas about how to create strong passwords. In this section, we will look at how passwords are used and what has been done to create strong passwords. This information will be used to help to adopt a happy medium between strong passwords and feasibility of ensuring that clients remember these passwords. This means creating strong passwords that are not so complex that they are written down.

Before we get into the password policy section, let us talk about the threats related to passwords themselves. Password cracking tools can attempt multiple password types until one is granted access. To understand how these work we need to talk about each kind detailing their functions. All password-attacking tools perform a guess on a password. This guess can be words that are part of a word list or dictionary. This is called a dictionary attack. If a password does not have a dictionary word, a brute force attack can be performed. A brute force attack will exhaust all possible combinations of characters. This test usually is time and resource restrictive, meaning without the right equipment and time span it may never find a correct password.

The first item that comes up when generically talking about passwords is the concept of factor identity. A password is considered a single factor authentication mechanism. The factor concept is made up of three main categories. These categories are defined as something you know such as a password, something you have such as a token or private key, and something you are such as fingerprint or retina scan. Using more than one factor reduces your risk dramatically. When deciding about adding additional factors to your password policy, remember to take into account that passwords are the least expensive means of authentication. Using another factor on top of passwords or in the place of passwords will increase security but it will also add expense and complexity. Tokens or biometrics along with the support costs of the added systems may prove to be more expensive then the added security provided by these systems.

When looking at what can be done with password policies we need to address password complexity. This complexity is an equalization technique used to find the strongest passwords without causing excessive password recovery issues. Using a complexity requirement will ensure that passwords are created strong from day one.

A complexity requirement involves the use of letters, numbers, or special characters. When developing this requirement a decision needs to be made as to how many of the three kinds of charters will be required. Requiring two or more of these complexity requirements will ensure that there are no passwords created from words found in a dictionary. Requiring all three will create a password that will require a long exhaustive brute force attack to reveal its context. The more complexity you add produces a negative effect. This effect is seen in an increase of password reset calls to the helpdesk and even worse the sticky note password on a workstation monitor. Requiring very complex passwords usually leads to users writing them down or forgetting them. A weak password is better than a strong one if the complexity makes users write it down and place it in front of their workstation.

Another interesting method used to create and remember passwords is the keyboard layout method. This is often taught to people with a high level of access to secret documents. It works rather well for governments due to the fact that users actually do not know the password; rather they are familiar with a certain set key layout. This gives a person the ability to take a polygraph and pass by honestly saying, "I do not know the password." This is because the user actually does not know the password; rather they know how to type it. To get into detail this method uses keys that are grouped together. This would be as easy as typing 4rfv5tgb. As you can see it has complexity and is somewhat random, yet if you were to type this on a keyboard you would instantly notice how easy it is to remember the key strokes and not the password itself.

One note about this type of password is most password attacking tools perform a check of closely grouped keys to find passwords like this. This may sound a little scary, although this step is often just before a brute force attack. With a brute force attack and time, any password can be recovered.

Access Policy

This section talks about who can access what wireless networks and with what devices. This section will address the immensely important issue of access and improper use of such access. To begin with, one must understand how the risk assessment addresses different levels of access. What risk rating was given to nonstandard, non-company-issued devices connecting to the wireless LAN? What risk rating if any, was given to standard company-approved and -deployed wireless devices? These ratings are often in direct relation to the current wireless network security architecture.

This section can also address what can and cannot access the network. Some devices such as smart phones with built-in Wi-Fi may be a real threat. There could be certain portions of a wireless network running mission critical application data. This is common in healthcare where patients' vital signs are monitored. In this example making sure there are no other unneeded or unwanted wireless devices that are transmitting on that network is critical. This policy should directly state what devices are allowed on what networks. This can be a list or a business rule stating a certain management level must approve the device before it can connect to a certain network. This section should clearly state what can happen to anyone who circumvents this policy and connects an unauthorized device.

Rogue Access Point Policy

Rogue access points can be placed by hackers or employees. In either case there needs to be a clear explanation of the consequences and actions team members must take if a rogue access point is identified. This could be as simple as to locate and remove any rogue access points or as dramatic as firing the person who installed it. Management needs to give input as to how forceful this portion should be in the written policy.

Guest Access Policy

Public or guest access is somewhat new to most security policies. Today hotspots or guest wireless networks are everywhere and most hospitals offer public wireless Internet access. Detailing this in the guest access policy will ensure that the guest access offered does not negatively impact the company. The policy should define what steps are needed to be able to provide guest access and what can be done to make sure the company's networks are secure.

The first portion of this section will address companies that would like to set up hotspots for customers. These networks should be separate from the corporate network and employ some security features to prevent a connection from jumping into the company network. This is where a policy stating certain high-level goals needs to be communicated. For example, the policy could state the hotspot connection needs to be physically disconnected from the company network and use a separate ISP. It could also specify using a certain security architecture, which lowers the risk and does not have the extra cost of a separate Internet connection. There are a few methods that are explored in this book on how to handle guest access. From the policy standpoint a high-level statement should be made that the security architecture approved follows.

Another way to look at it is to disallow wireless access for employees. This may make some employee mad and lead to rogue access points. One way to fix that is to design the hotspot network for customer use and company use. This is a secure approach to allow for bring your own device (BYOD). For company users they could use a VPN to connect to the company resources. This would be the same case if they were using another hotspot down the road. This approach ensures the user experience is the same. If this approach is used the policy only needs to address employee wireless access regardless of where the access is originating from. This is since access could be through a VPN onsite or through the same VPN at a public hotspot. Whatever route a policy may go, it is important to understand that there is significant risk involved in running a hotspot or guest network. This risk can be reduced by performing a risk assessment and developing a guest access policy well before the technical matters are defined.

Remote WLAN Access Policy

In this section we are going to address the issue of using hotspots to connect to your company network. Most hotspots have little or no security at all. Making sure a hacker does not steal or piggyback off one of your remote user's laptop is very important in protecting the organization and lowering risk. When a proper risk assessment is performed, a number of risks may come out that do not directly relate to the wireless network. Some technologies such as e-mail retrieval and remote management have common flaws. These flaws are considered acceptable in the situation where someone is connected to a wire that is connected to the Internet. The Internet may be a wild place where your traffic can float around the globe, although the only likely candidates that would be able to sniff your network traffic are employees of service providers. That risk is often considered acceptable when these insecure technologies were evaluated and deployed. Now with wireless anyone can see the traffic; different rules and re-evaluations need to be done regarding these insecure protocols.

The last portion has to deal with someone who would steal company information off a wireless device. The next section of physical security will address the theft of the device itself. In this section we are going to address the data separately since there are many ways of stealing the data off a device without having to steal the device itself. If we look at using wireless technologies such as bluetooth or Wi-Fi for hacking many ways exist to compromise wireless devices. Protecting the data on the device is extremely important. Many smart phones carry contacts and sales information that is critical to companies. If this data fell into the wrong hands, a business could quickly be out of business. To protect this data some steps need to be set forth in the policy. One might be able to use password protection on all wireless end devices. Another would be to disallow creation of types of communication technologies not considered secure. However this is done, careful attention needs to be given to this section because in most cases the data on these devices are the most important asset of a company compared to the inexpensive cost of the hardware.

Physical Security

The physical security has its own unique aspects. One of them deals with the wireless network itself. How many times have you seen a wireless network? The next time you are in any retail store look around and you will see them. You will see the antennas hanging; sometimes you will see the access points themselves. In this case, the access points and antennas are located well out of reach. In areas like office buildings and hospitals, often the access points are located in the hallways, where everyone can see them and touch them. This is a physical security risk and controls need to be incorporated in this policy section. Another common physical security issue is that of the end devices themselves. These devices are small and often forgotten or stolen. Making sure they are safe from thieves is an important part of this policy. In this section, we will look at each of the risks that directly relate to wireless physical security and how certain policy steps can mitigate or prevent them altogether.

The first risk comes from stealing the access point themselves. Some common commercial access points can range in cost from $100 to over $1000. Keeping them safe is important. Today almost everyone is using wireless networks at home. This means the threat of someone walking off with a wireless access point is very likely. This risk is even greater if the facility is open to the public like a hospital. To protect against this, certain rules need to be set. These rules can be set in this section of the policy. Making it a policy to use enclosures to house access points can reduce this risk. Another approach would be to require the access points to be mounted out of sight. This would be as simple as putting them above a ceiling tile. Their antennas may still need to be located below the tile but they are less likely to draw attention than an access point with its flashing lights attached to a wall within reach.

The next portion of the policy relates to wireless end devices. As stated above these devices are easily stolen or lost. To help prevent this some steps need to be taken. For example, do not allow company assets to be left unattended. One step that can be taken to mitigate this risk would be to use software that can send cellular signals to these devices. These signals can erase the device and lock it from use. This technology is very helpful if a wireless device is stolen or lost.

Education is the key to preventing this from happening. If proper steps and technology are used, correctly safeguarding these devices will lower the risk of having them lost or stolen.

Wireless Monitoring and Security Incident Response

In environments like hospitals there are compliance regulations that impose some type of monitoring and resulting actions. This policy section would address how to set up such monitoring, for example, how to configure a wireless intrusion detection system. This should address a baseline configuration and monitor set, a logging architecture, and integration into any already established security or network operation center management technologies or processes.

HIPAA and Wi-Fi

On August 21, 1996, the U.S. Congress passed the Health Insurance Portability and Accountability Act (HIPAA). The act has many parts; the major items that relate to technology and especially wireless local area network technology will be discussed here. After reading this section you will have a full understanding of how the law applies to wireless local area network technology.

One of the main mandates of HIPAA is to protect personal health information or what is commonly referred to as PHI. This is also call ePHI when the information is in an electronic format. The HIPAA mandate applies to covered entities. A "covered entity" is generally any entity that acts as a healthcare provider, healthcare clearinghouse, or health plan. Doctors, clinics, psychologists, dentists, chiropractors, nursing homes, and pharmacies are generally all considered healthcare providers under HIPAA. One of the test criteria used to see if an entity is or is not a covered entity under the healthcare provider category is if they furnish, bill, or receive payment for healthcare in the normal course of business. Entries such as health insurance companies, health maintenance organizations (HMOs), company health plans, Medicare, Medicaid, and the military and veterans' healthcare program providers fall under the health plan category. The main test for this type is if they provide or pay for the cost of medical care. The last type of "covered entity" is healthcare clearinghouses. The test for these types of entities is if they process, or facilitate the processing of, health information.

The main part of HIPAA that we need to be concerned about relates to the HIPAA Security Rule. The Security Rule is located at 45 CFR Part 160 and Subparts A and C of Part 164. This rule sets standards to protect individuals' electronic personal health information that is created, received, used, or maintained by a covered entity. The Security Rule has three parts: Administrative Safeguards, Physical Safeguards, and Technical Safeguards.

The Security Rule defines administrative safeguards as "administrative actions, and policies and procedures, to manage the selection, development, implementation, and maintenance of security measures to protect electronic protected health information and to manage the conduct of the covered entity's workforce in relation to the protection of that information." The Security Rule defines physical safeguards as "physical measures, policies, and procedures to protect a covered entity's electronic information systems and related buildings and equipment, from natural and environmental hazards, and unauthorized intrusion." The Security Rule defines technical safeguards as "the technology and the policy and procedures for its use that protect electronic protected health information and control access to it."

Further information about HIPAA and how to apply the security rules can be obtained from the National Institute of Standards and Technology (NIST) Special Publications. NIST is a federal agency that sets computer security standards for the federal government and publishes reports on topics related to IT security. Some of these reports provide more technical details about how to meet certain HIPAA Security Rules. The common ones are listed below.

- NIST Special Publication 800-30: Risk Management Guide for Information Technology Systems
- NIST Special Publication 800-52: Guidelines for the Selection and Use of Transport Layer Security (TLS) Implementations
- NIST Special Publication 800-66: An Introductory Resource Guide for Implementing the HIPAA Security Rule
- NIST Special Publication 800-77: Guide to IPsec VPNs
- NIST Special Publication 800-88: Computer Security

- NIST Special Publication 800-111: Guide to Storage Encryption Technologies for End User Devices
- NIST Special Publication 800-113: Guide to SSL VPNs
- Federal Information Processing Standards Publication 140-2

Now that we have a full understanding about HIPAA and what items relate to technology we need to take a more detailed look at how the HIPAA Security Rule relates to wireless local area networks. There are a few topics that are not specific to wireless local area networks but are still major requirements of HIPAA; they need to be addressed at a much higher level within an information technology function. An example of this is the risk analysis rule under § 164.308(a)(1)(ii)(A). You need a risk assessment, although that should take place and account for testing anywhere within the entity where PHI might be transmitted. In our case this means since PHI could be transmitted over a wireless network the network itself should be taken into account during this analysis.

As we dive into HIPAA and its relation to wireless local area networking we must look at different sections of the HIPAA Security Rule. The first subsection we will explore is also the most relevant as it addresses technical safeguards. HIPAA defines Security Rule Technical Safeguards in § 164.304 as "the technology and the policy and procedures for its use that protect electronic protected health information and control access to it." We will explore each area of these various technical safeguards and how they relate to wireless local area networking.

The first relevant safeguard is § 164.312(a)(1) Access Control. The legal definition of this falls back onto another section of HIPAA that relates to administrative safeguards. How these two relate to each other is that the administrative part defines the policy around whom and the technical part enforces such a policy with the how. Here is the official definition: "Implement technical policies and procedures for electronic information systems that maintain electronic protected health information to allow access only to those persons or software programs that have been granted access rights as specified in § 164.308(a)(4)[Information Access Management]."

The access control safeguard is further refined into four areas, the first of which is § 164.312(a)(2)(i) Unique User Identification. The official definition is: "assign a unique name and/or number for identifying and tracking user identity." This item relates to wireless local area networking depending on how your wireless security is designed. Wireless security can be set up using WPA2 enterprise with a backend authentication system that is an organization directory such as Microsoft Active Directory (AD), Lightweight Directory Access Protocol (LDAP), or a number of other directory systems. This entire item could be classified out of scope from the wireless side if such access controls are addressed on the systems that house the PHI-related information. Why it is included is because many wireless networks have various logical wireless networks with varying levels of access based on user rights. So users with needs for PHI access connect to one logical wireless networking that is connected to a VLAN where PHI exists or can be accessed. Other users who need no PHI access connect to another logical wireless network that is connected to a separate VLAN without PHI access.

The next subsection of the access control safeguard is an emergency access procedure § 164.312(a)(2)(iii). Normally this would not be related to wireless local area networking, although a wireless network could be used as a secondary method to access ePHI in the event the primary method is unavailable. This would not be a normal situation, although it could be used to satisfy the requirement in unique situations.

Another subsection of the access control safeguard is automatic logoff § 164.312(a)(2)(iii). This is normal and should be addressed on the application or data store location of where any ePHI resides. A wireless local area network could allow for you to apply this, although there are many better places to address this matter, like on the actual end device or as stated where the actual ePHI resides.

The last subsection of the access control safeguard is the most relevant to wireless local area networking. It is Encryption and Decryption § 164.312(a)(2)(iv). This section has the grayest area when it comes to information technology departments' and legal departments' viewpoints. The reasoning for this gray area is the reasonable and appropriate words used within. The official definition is: "Implement a mechanism to encrypt and decrypt electronic protected health information." In relation to any gray area as far as the wireless network is considered, in almost all cases you must encrypt a wireless network as an unencrypted wireless network is not considered reasonable or appropriate for any business use.

This section is another area where the actual PHI could be protected and encrypted by a number of encryption methods outside the wireless network. While that might address this single HIPAA requirement, having an unencrypted wireless network is a major problem no matter what type of data is being transferred over it. In recent years the type of encryption has even came into account as many other regulations outside HIPAA, such as those for the payment card industry (PCI), have discounted various types of wireless encryption methods as violations of that standard. WEP is an example of this—although WEP provides encryption it is widely considered insecure. With this said, you can see the gray area as this is encryption but how it relates to reasonable and appropriate is debatable. This concludes the various subsections of the access control technical safeguard.

The next area of the technical safeguard is § 164.312(b) Audit Controls. The official definition is: "implement hardware, software, and/or procedural mechanisms that record and examine activity in information systems that contain or use electronic protected health information." This is another somewhat gray area in the sense that it falls under "reasonable and appropriate." Almost all technology systems provide some form of auditing and logging. Like access control this is driven more from the output of other policy items or the results of your risk analysis. One area inside HIPAA that relates to this is how you address the § 164.308(a)(1)(ii)(D) for Information System Activity Review. This part is outside the scope for Wi-Fi, although depending on how this was addressed the results could highly influence this section.

An important factor to consider with this is that the safeguard does not identify any type of data that should or must be audited. A good point of reference to consider when selecting what to audit is to consider what audit data could be used to determine a security violation. Another important factor to consider is if the safeguard does not identify the frequency or review period.

The next section of the technical safeguard is § 164.312(c) Integrity. The official definition is: "the property that data or information have not been altered or destroyed in an unauthorized manner." This item is about ensuring that policies and procedures exist to protect ePHI from improper alteration or destruction. This includes both technical and nontechnical methods. These types of controls should exist outside the WLAN and be related to the ePHI itself. The WLAN has various ways to uphold the integrity of all communications across the airwaves depending on how the security is set up.

The next section is called Mechanism to authenticate electronic protected health information (A) - § 164.312(c)(2). The official definition is: "Implement electronic mechanisms to corroborate that electronic protected health information has not been altered or destroyed in an unauthorized manner." How this is assessed is based off the risk analysis that should have been performed. This section is a validation of technical integrity controls such as checksums and digital signatures.

The next section is called: Person or Entity Authentication § 164.312(d). The official definition is: "Implement procedures to verify that a person or entity seeking access to electronic protected health information is the one claimed." This item can be addressed by various wireless security methodologies as they leverage various types of user-based authentication. Mostly this is handled nearest to where the actual ePHI is located. An example of this would be the application that holds the ePHI would have user-based authentication. This would supersede the network which allowed access to the application.

The next section is called: Transmission Security § 164.312(e)(1). The official definition is: "Implement technical security measures to guard against unauthorized access to electronic protected health information that is being transmitted over an electronic communications network." Almost all of the wireless security methods address this item. Over time some of the weaker types of wireless security such as WEP have been deemed so insecure that it will not qualify for any type of transmission security. This is a gray area as some standards such as those for the PCI actually specify WEP as unacceptable. HIPAA does not get into that amount of detail about inadequate encryption.

Part of the transmission security section also covers Integrity Controls (A) - § 164.312(e)(2)(i). The official definition of this is: "Implement security measures to ensure that electronically transmitted electronic protected health information is not improperly modified without detection until disposed of." This relates to PHI being modified while in transit. When evaluating this some thoughts to consider are if there is ability to modify PHI while in transit and what integrity controls exists today to prevent such action.

The last section within the transmission security addresses Encryption (A) - § 164.312(e)(2)(ii). The official definition of this is: "Implement a mechanism to encrypt electronic protected health information whenever deemed appropriate." This section is only speaking about the confidentiality of information while being transferred. This section does not address any technical items related to how. In the area of wireless any PHI traveling across an encrypted wireless link is compliant as long as the encryption is not poor like WEP. Officially WEP might be considered acceptable, although it should not as it is commonly deemed as a poor wireless security choice. The real concern with this item is the guest network or any type of bring your own device. A lot of these secondary networks have no encryption and could easily be accidently used to transfer some sort of PHI and result in a violation.

5
WIRELESS GUEST SERVICES

The concept of wireless guest access has been around since the early days of 802.11b. It was quickly realized that the convenience of not having to plug into a wire to access the Internet enabled a "why can't I work here?" mindset. At the start of wireless guest access we saw mostly hospitality use cases for guest access. Like the coffee shop or hotel where travels wanted to connect to the Internet and work or play with technology where and how they wanted. Over time this changed to a question of why do I need to plug into the network at all?

Once wireless access was granted by companies for employee use the next logical request resulted; why can't our trusted partners use the wireless? Why do they have to leave, go to the closest coffee shop to download a file or the latest copy of the presentation, or check for that important e-mail. We have wireless in our company why can't they use it? The common technology department answers resulted. The most common of these were that for security reasons we cannot intermix protected and unprotected devices or that there were no documented policies or standards on the matter of allowing non-employees to connect to a company network. The question of who would accept the legal risks this could bring was just one of the many questions surrounding how wireless could or should not be used by third parties.

Smart companies started listening to what their workers wanted and realized that just as wireless increased their own workers' productivity it could also increase the productivity of their trusted partners. As technology pushed more toward a cloud-based model where most applications require an Internet connection to operate, the need to allow various third parties to access the Internet from anywhere is rapidly increasing. This is fast becoming a standard practice and many visiting customers or partners have started to expect, demand, or request some method to maintain an Internet connection while visiting.

Within healthcare wireless networks, guest access is a major hurdle to address. Both the hospitality and third-party business enablement use cases are present. Add to that a highly regulated technology environment where data loss or accidental disclosure has enormous consequences and you start to understand the challenge of placing guest access in the healthcare environment. Even if these issues have been addressed there remain the pockets of wireless that could result in the loss of human life if impeded or accidentally disrupted by a third party or guest.

This chapter sheds some light on various methods of providing a guest wireless network. The main difference between pure guests and trusted third parties is how the network allows access and if some guests will be required to pay for their access or not. In this chapter we explore a few methods that have been used to address these problems.

Sponsored, Open Access, and Self-Enrollment

Most guest networks are sponsored, self-service, or completely open. There are many reasons for setting up a guest wireless network one way or another.

Sponsored Guest Access

With sponsored access the wireless network requires some form of identification information to be presented by the guest. What this information is and how it is presented varies widely between the various solutions that exist in the market. There are situations where a person must create an account for a guest and collect certain elements of information before doing so. A lot of what this information is focuses around company policies stemming from legal requirements and not so much based on the technical options the various solutions allow. For example, the information you may request could be considered patient information and might be subject to HIPAA security controls. Various national governmental bodies might require certain types of information to be captured before providing any type of Internet access.

In reference to security the guest network in a high-security environment might be a protected encrypted network requiring privileged information to be provided before allowing the user to access the network. There are solutions on the market where a person must be provisioned for connecting to that network before anything can happen. Some solutions allow for the creation of a X.509 PKI certificate for use by the guest for the duration of their stay. Some solutions fall short of the high-security approach by merely allowing the guest a MAC address into a table of allowable devices. This means a somewhat sophisticated hacker could easily masquerade as the sponsored guest in environments where simple security such as MAC address filtering is imposed. Most hospitality types of guest access have the delicate balance of ensuring the network is easy to use and keeping the security at an adequate level. Some businesses have started with the sponsored type of guest network only to have their corporate counsel advise against collecting information as it introduces the risk of not collecting the right information. Some regulations or laws may require a sponsored type of access.

Self-Enrollment Guest Access

There are solutions where identifying information is asked of guests when they attempt to access the network via a splash screen. This information has implications for HIPAA compliance if the information requested is considered PHI. Also depending on how this information is verified or if it is even verified, there is a lot of risk for false information if such information is required by law or corporate policy. For example, many existing solutions ask you for your name, e-mail address, and phone number. None of this information is verified outside ensuring something is inserted into the fields.

Some of the more secure solutions will send your access information or credentials to the e-mail address you provide, or send a text message to your phone. As you can imagine some of these methods could be a problem. For example, how do you check an e-mail for information that will give you access to the Internet before you have access? This is an example of features before common sense. The text or SMS messaging solution is commonly used because a cell phone is normally traceable back to a person should a legal matter result. While this might sound like the best approach thus far, the problem mounts when you attempt to allow any guest a connection with this approach. Security will always slightly raise complexity and if the goal is convenience and service satisfaction such as wireless for hospitality purposes, you might find your new service is increasing dissatisfaction or costing more than expected in support charges since the security is considered burdensome.

Open Access

More and more guest systems exist with completely open access. There are many levels to this as open access only refers to the fact that you can get onto the network without impediment. Once on the network most of these have some sort of basic URL filtering so as to prevent the viewing of inappropriate content. Outside the United States there are many laws that require the tracking of Internet users beyond what is required in the United States. Some areas of the world would not be allowed to have open access. If you are looking to create a globally consistent guest network across a global enterprise you may want to consider sponsored access instead of open access.

Open access is a good option where the WLAN infrastructure does not support the necessary features to implement a registration page or even a simple terms-of-use acceptance page. It is strongly advised that open guest access be rate limited. Many mobile devices have the option to automatically connect to Wi-Fi networks. This will likely result in many of these devices passively leaching onto the guest network to perform application updates and e-mail downloads. While there is nothing inherently wrong with providing a service like this, it provides little to no value to the unaware end user but may cause the hospital to incur additional costs. Wi-Fi leeches are difficult to track and prevent.

Captive Portal Page Types   A captive portal could be anything from a simple splash page to self-registration with a billing option. Having a captive portal can provide an opportunity to interact with patients and guests. It could provide detailed information about the hospital or direct the patients and guests to the organization's mobile applications. There are even possibilities to provide advertising. Although this sounds great the reality is that most people are very put off by any additional steps needed to get connected to the Internet. However, a well-designed captive portal (CP) page can provide the features the provider needs without irritating the guest. Before diving into your Wi-Fi vendor's captive portal design tool, it will help to take a look at the various types of CP pages.

No Registration Splash Page   A no registration splash page requires the least intervention from the guest user. This type of page can range from a simple text only "Welcome" page with a timeout to a page with professional graphics, browser detection, and terms and conditions written by the legal department. An acceptable use policy (AUP) is strongly recommended. This should provide at least an initial defense against any damages that are sourced from your guest network. If a virus-infected computer is on your guest network spamming G-mail accounts, Google may block the guest Wi-Fi public IP address from accessing G-mail or blacklist it from Google altogether. While your organization may have legal recourse regardless of whether you enlist an AUP, the AUP will provide clear breach of the agreement. If this IP address is a shared corporate network, this could result in monetary damages for the health organization. One important thing to note is a splash page has an inherent anonymity. The Wi-Fi network only knows a malicious device as an IP or MAC address. In order for the guest device to become less anonymous you must begin to look towards registration captive portal pages.

One clear advantage to using a captive portal page is that they provide a speed bump for many data-hungry mobile devices. This will help prevent unintentional data use. Most mobile devices have the ability to automatically connect to any nearby Wi-Fi network and begin using it for background data transfers. In addition, most devices prefer the Wi-Fi network over the 3G/4G connection and will use this first. As a result, a guest network without a splash page could result in a substantial amount of bandwidth performing things like App downloads and updates to things like the weather forecast and stock prices that are not even being used or even by a guest at the current time. Multitenant buildings and medical centers in shopping plazas are typically in close proximity to nonpatients and nonpatient guests. While the convenience of not having a splash page resonates well with guests, it may be necessary to weigh costs of this additional bandwidth consumption against the ease of use.

Self-Registration   An organization may find itself looking to decrease the anonymity of guests. While this may seem draconian, the intent is hopefully to improve the experience for patients and patient guests. Since Wi-Fi is a finite resource it needs to be managed. Using a captive portal with self-registration is a great way to limit access to that finite resource while maintaining ease of use. Self-registration can come in many forms. A simple self-registration page may require a guest to enter his or her e-mail address in a web form on the captive portal. Something more complicated might require the guest to provide a cell phone number that an access code would be sent to via SMS text message.

How the registration occurs depends on the available options of the Wi-Fi infrastructure. Options like SMS text messaging are likely to be provided by a third-party company with some sort of integration into the Wi-Fi infrastructure such as RADIUS. This level of sophistication may not be for everyone. For starters, third-party integrators will increase the cost of a guest Wi-Fi installation. It is likely that this will be reserved for larger deployments, and in multitenant buildings. However, cloud-based solutions make it possible to implement this technology regardless of the scale of the deployment. For smaller Wi-Fi installations, the organization may phase in more sophisticated solutions over time. So, if you are just beginning it may make sense to begin with features that are available out of the box.

Manual Registration   In a manual registration scenario, one or more persons are guest access administrators. When a guest needs to get access to Wi-Fi he or she will have to request it from the administrator. Often this can be a hospitality representative or in office environments it can be the person at the lobby front desk. In healthcare, this can be a nurse or a helpdesk worker. Either way, this functionality is usually available out of the box in WLAN equipment. A guest administrator login can be created for a nurse or anyone who would receive a request for access. The guest administrator can then print out or even write down the guest user's login information and provide this to the patient or guest. The patient or guest then would enter this login information into a login section of a captive portal page. Manual registration is a good starting point to use before moving to self-registration as it can help to understand the guest user's requirements for access. Manual registration is the ultimate speed bump when it comes to guest Wi-Fi. This can help really limit who gets access to the guest network. This may prove especially valuable in deterring employees from using their personal devices on the guest network. On the other hand, this can really annoy patients and patient guests. Since this process relies on the intervention of employees, you will need to ensure that there is enough personnel to support the guest requests.

Sponsored Registration Sponsored registration is similar to self-registration; however, the guest identifies an employee's e-mail address. The employee would receive an e-mail in response to which they would either acknowledge or deny the guest access to the Internet. This scenario is more common in non-public areas but could be used in a hospital setting if guest access was not offered to all guests, for example.

Guest access is a continually evolving problem to solve. New innovations to this problem abound. As a result, it is important to always define, measure, and possibly redesign the solutions to best meet the needs of guests. What works today may not work tomorrow. This perhaps is one of the most challenging aspects of guest access. The explosion of mobile devices is a testament to this.

Each guest Wi-Fi solution will have its own pros and cons to consider. Regardless of the selected solution, it is likely that employees will want to use their personal devices on the guest network. This is primarily because of the ease of use. The majority of these devices are smart phones and tablets. It is fascinating to watch as these devices seem to grow in numbers with no slowing in sight. This team of authors has experienced firsthand something we call the Santa Claus effect. As the name implies, it occurs after Christmas has come and gone, usually around the first of the year. This could result in a 20 percent increase of guest client counts as many employees have brought in their new tablet or smart phone they received for Christmas or other holiday. It would be great if Santa's elves would give us all a heads up before Christmas morning. Then it might not be necessary to make an emergency change the following week to expand DHCP scopes. Or, better yet we can give this phenomenon a four-letter acronym and part of a book chapter on guest wireless. The subject of Bring Your Own Device (BYOD) can evoke emotions, both good and bad.

Supporting Infrastructure

Regardless of the solution you choose for providing guest access, it is very likely you will minimally need DHCP and DNS services. Additionally, you may need RADIUS and Web hosting services for captive portal pages and authentication. All WLAN vendors have some type of built-in splash page or even self-registration page available. Guest location tracking may require another appliance or software to be loaded onto a server. When thinking about how to achieve guest access make sure that proper attention is given to understanding the network services required.

Revenue Generation

In the United States, guest Wi-Fi revenue generation in a hospital, medical center, or even clinic is a bit of a taboo for most organizations. Many will say that Wi-Fi should be free as in the air we breathe. However, in other countries guest Wi-Fi is viewed as another potential entertainment revenue source and visitors are more accepting that one has to pay for access. Others suggest this just leads to 1 percent of the users ruining it for the rest. One common practice of guest Wi-Fi is implementing a CIR or Committed Information Rate. This is a fancy way of saying you are giving guests a slower connection speed. The CIR serves as a speed bump for the guest users. A CIR of 512 K would be a moderately conservative policy to implement. However, this would permit a reasonable experience for web browsing and messaging. However, if you compare this bandwidth to a 3G, EVDO, or 4G/LTE connection, it may change your opinion on how good of a connection you are provided. In a pay-for-performance model the CIR would be much higher or eliminated. Another option would be a graduated CIR. An example of this would provide 1 Mb of data at full speed. After that point a CIR of something like 256 Kbps will kick in. The CIR will provide free access with limits. However, some guests may demand more bandwidth. Suppose an individual would like to watch their DVR back home. It is somewhat unreasonable to believe that an individual should be able to watch a TV program in HD over a free Internet connection. I suspect most people would not expect this to work well over their 3G connection. However, Wi-Fi does have a certain mystique that home users find limitless. Which is a good point to bring up. I have found that the average guest user would *like* their experience to be as good as their home wireless and *expect* it to be better than their cellular data connection. Unfortunately for healthcare providers, this puts the bandwidth expectancy north of the conservative 512 Kbps CIR I mentioned earlier. To that point, it would make sense for the CIR to at least match what is possible from a cellular connection in the area.

Bring Your Own Device (BYOD)

Wi-Fi networks were born out of application-specific devices. The first WLAN this author was fortunate enough to implement utilized the 802.11 protocol. This predates even 802.11b and is incompatible with the current WLANs, not to mention it was connected to a token ring network. But never mind that, in those early days of wireless things like handheld scanners were the primary driver for mobility. And this was a good fit as the data throughput and latency requirements were low. Now that first network from earlier could only sustain 1 Mbps, which after overhead was not much better than a dial-up modem. As we fast-forward to the present technology, we now see connectivity speeds up to 450 Mbps and there are new standards on the horizon with even higher throughput. All these increases in throughput have reached a point where we are seeing diminishing returns on that bandwidth. In fact, most mobile devices will not be able to sustain actual throughput in excess of 35 to 40 Mbps. Also, keep in mind that most mobile devices are receiving the majority of their content from the Internet. So, the reality is there is more throughput available on an AP than we can currently utilize. This fact and two companies, Apple and Google to be exact, have created a perfect storm for a mobile explosion. This explosion has left many corporate IT departments trying to catch up. Early adopters to Wi-Fi found themselves defining what devices would be used and how they would be configured. But something called "consumerization" occurred. This is a very new concept. In fact, the word consumerization was only just added to the *Oxford English Dictionary* in 2009. History has shown us that laptops were expensive and mostly used by employees who worked outside of their offices. PDAs and smart phones were for geeks mostly.

There has been a great paradigm shift in what technology can be used for. Companies like RIM, Apple, and Google began to make geeky a fashion statement. This added fuel to the fire and the explosion began. Average people began to have an option on how they would like to consume their daily intake of information. This paradigm shift to IT departments has largely been viewed as a disruptive revolution. This is IT speak for a major annoyance. However, it could be argued that this is a logical evolution. Personal preference is simply human nature. So perhaps IT departments should take a deep breath and embrace this brave new world. But keep in mind that actually accomplishing this massive feat is another story. After all, healthcare is a government-regulated industry. So, how on earth does one provide choice and personal preference in the way that data is consumed while at the same time protecting the data and meeting government regulation and compliance? When stated that way, the task is less forward thinking and more arduous. And, it is exactly that.

The difficult task with any new initiative is to decide where to begin and then actually take that first step. Deciding where to begin and how much to bite off will be the first step of a BYOD program. The outcome of this initiative will be a written BYOD policy, and some technology to support this policy. It is unlikely that an organization will be able to specifically accommodate every device that is manufactured from now till 2050 with a policy written in 2013. In line with that, adaptation should be the core of any policy. At the time of this writing, Apple is enjoying the lion's share of acceptance from enterprises. However, this could easily shift to Google's offerings or even a newcomer, maybe Facebook.

During the discovery phase, tools play an incredibly important role in understanding the device population and business drivers. It will be important to gather as much information about the device population and how they interface with applications. The Wi-Fi infrastructure will likely have reporting capabilities. This is a great place to look for this information. Additionally, Network Access Control (NAC) products have auditing capabilities. This will likely provide information about not only what types of devices already exist on the network but also what applications are being used. Alternatives to NAC are packet capture and analysis tools, which come in all shapes and sizes.

Once an organization has documented what devices the user population is currently using, it will need to identify how to securely and effectively connect the device to the WLAN. This might require quite a bit of trial and error. One problem will be moving away from solutions that have worked in the past or on other devices but will not work for current and future devices. For example, consider a health system that implemented a WLAN with WPA2 encryption and Protected EAP (PEAP) for authentication. This was a great solution for the organization during the laptop boom as desktops were replaced with laptops. Pre-BYOD most employees were only issued one laptop that was managed by the corporate IT organization. Fast-forward to now, and we find the average person carries three to five Wi-Fi-capable devices, of which only one or two might be managed. This has added some complexity to simple operational procedures such as password changes. Many smart phones offer their owner the option to cache passwords for things like e-mail, web forms, and the Wi-Fi settings. When a password expires it must be changed. When this act must be synchronized across three to five devices and possibly on multiple applications on each device, it is likely the helpdesk will get a lot of calls to reset passwords. This used to be a snap since laptops were configured to use the credentials from a Windows or Novell login. There was one place to change your password and only one place to login. Smart phones and tablets from Apple and Google do not have a login screen like laptops. Therefore, they will need to have the WLAN authentication credentials provided from elsewhere. This will likely require the devices to cache the password. When it is time to change that password, every device will need to change or delete the cached password. To make a long story short, PEAP was designed to primarily authorized Windows laptops. In fact, this was a joint venture between Microsoft and Cisco over 10 years ago. That relationship was later strained and Cisco developed its own authentication solution called EAP-FAST. What worked in the past will not likely work for newer devices. Once again, it is important that a BYOD policy is adaptive and even more important that the WLAN is equally adaptive.

Mobile device management (MDM) is a key component to BYOD. The primary role of this element is to provision and audit mobile devices such as phones and tablets. Provisioning mobile devices is not a new concept. RIM, who is the maker of the once popular Blackberry, pioneered MDM. RIM sold a solution called the Blackberry Enterprise Server. One of the key features of this solution was to provision every Blackberry phone in an enterprise with features such as password lock policy and even custom applications. A lost phone could even be wiped clean remotely. There are many providers of MDM solutions on the market. One thing that they are likely to all have in common is that there is a cost associated with every device that is connected to them. This one factor may greatly change the scope of a BYOD policy. Not every organization will be enthusiastic about paying for every device that connects to their corporate WLAN. For this reason, it might make sense to qualify what devices are covered by corporate MDM. It also may make sense to charge for devices that are not used in the line of business. As of this writing, the most complete and common MDMs are stand-alone systems. It is the opinion of the authors that there will be many mergers and acquisitions that will lead to these systems being integrated into larger network equipment and software vendors' product lines. Regardless of who is providing this functionality, it is a critical part of a successful BYOD program.

In the late 1990s and early 2000s one of the jobs in highest demand was a Windows server administrator. Windows administration involved creating users and groups then applying privileges to both of them. In order for the privileges to be applied to a user they had to login to a server or Windows domain. This allowed the purchasing department to reach the procurement application and human resource to access the payroll files. While this still happens even today, there is a new complexity that has been added with mobile devices. When a mobile device connects to a WLAN it is placed alongside the corporate managed laptops. However, there is nothing that is either allowing or preventing access to local resources such as the applications and files mentioned prior. While it may not be of too great a concern that the mobile devices are able to route to important servers, the inverse may be a big problem. Smart phones and tablets are not the best tools to access things like files servers and Microsoft Windows-based applications. There are of course work-arounds such as a Virtual Desktop Infrastructure (VDI). However, these devices are fundamentally different from PCs. There is no Windows domain to manage these devices. At the same time this is an intentional change in how people are choosing to work. Nonetheless, due to government regulation and cyber wars being waged in the United States and abroad there is a certain need for a traffic cop to sometimes intervene. This traffic cop has a name which is NAC or Network Access Control. This too is a critical component to BYOD for the reason just stated. NAC is quite possibly one of the vaguest collections of technology. And, depending on which NAC vendor you are speaking with, their definitions of what this technology can do can differ markedly. Every NAC solution will support a number of standard technologies along with patented technologies. Among those standard technologies will be:

- SNMP polling and trap collection
- Dynamic VLAN assignment
- DHCP snooping
- Kerberos snooping
- Client agents

SNMP polling and trap collection will mostly involve the wireless LAN infrastructure. SNMP is one of the most basic methods of collecting data about the infrastructure and clients. It is important that the management information base (MIB) of the WLAN infrastructure is very complete to this end. This uses something called ASN.1 to organize the data about a WLAN or any piece of network equipment, for that matter, in a hierarchical manner. Ensuring that the MIB for the WLAN has as many variables as possible will be helpful. Network access control can use this information to make decisions about how to route a client's data. One specific method of routing this traffic is through the use of dynamic VLAN's. Dynamic VLAN assignment may apply more to wired access than wireless access but some vendors implement this functionality into their access points. Dynamic VLAN assignment is discussed in detail in RFC 3580, which is the standard for 802.1X RADIUS protocol. DHCP snooping can provide a lot of insight into the clients that connect to a WLAN. For instance things like the client hostname, MAC address, and even operating system type and version can be collected for the purpose of fingerprinting. Fingerprinting is not only a method for applying access control to any device, but also it can provide granularity in reporting and auditing tools. In addition to these, Kerberos is a method of providing users and devices access to network resources such as servers. Kerberos was developed at MIT under Project Athena. Companies such as Microsoft have integrated Kerberos into server 2008 and 2012. Kerberos works off a principle of granting tickets to allow access and then revoking or renewing it. Inspecting this traffic can help understand what devices are being used to access patient health information (PHI) and other data. I should note that this functionality needs to be either integrated into the wireless infrastructure or positioned inline with a gateway device. Alternatively, this may be positioned out of band using something called a tap. Taps work by making a copy of all the traffic they see and sending them to a data collection device like a NAC gateway. Lastly, client agents are one of the most powerful pieces of the NAC puzzle. It is easy to guess that the client agent is a piece of software which is loaded on the client devices for the purpose of interrogation. They can come in at least two configurations: lightweight and stand-alone. Many providers also have a third flavor that is a lightweight agent that uninstalls after inspection. Agents can be an annoyance as they can interfere with the user's work by preventing them from using the device while it scans. Agents will scan for a host of attributes about a client such as whether it is up to date on software patches and anti-virus definitions. This can also be performed in a noninvasive manner or one time only during and on boarding. It is important to keep in mind that agents may prevent users from working during device scans. Therefore, it is important to completely vet all operational procedures associated with agents prior to their deployment.

Device certificates are a cornerstone of a BYOD solution. Usernames and passwords have proven to be problematic in modern networks. WLAN origins had a one-to-one correlation of users to devices. That correlation has changed to a one-to-many correlation. Password changes often require the synchronization of cached passwords and changing them on multiple devices. This is where certificates can provide some assistance. Password-based authentication schemes are easy to implement and deploy but painful to users. Certificate-based authentication, however, provides a better experience to the end user at the expense of a more complicated deployment. The process for managing user passwords will likely be very similar to managing certificates. However, there are more nuances to the creation and storing of certificates in the enterprise. With that said, the BYOD eruption has helped to introduce more tools to help manage the lifecycle of a Public Key Infrastructure (PKI).

SCEP

Simple Certificate Enrollment Protocol (SCEP) is a not a new technology but is still in IETF draft form. SCEP aids in the lifecycle management of certificates.[1] SCEP has features for certificate enrollment, renewal, certificate authority retrieval, and certificate rollover. Security measures are often described as being something you know, something you are, or something you have. Something you know is often a password, something you are can be a biometric sensor, and something you have is often a certificate. The password lifecycle usually begins with an initial password after an account is created. Thereafter, it is stuck in a continual routine change cycle until the account is disabled or removed. With certificate lifecycle, instead of continual password changes there are certificate renewals. These are usually halfway through the life of a certificate. Microsoft uses default values of five years typically. These events are less frequent and should not require any user intervention.

Endnotes

1. SCEP can be reviewed at http://www.ietf.org/proceedings/69/slides/pkix -3.pdf.

6

MOBILE MEDICAL DEVICES

Wireless technology has played a significant role in reshaping health-care over the last two decades. Wi-Fi began to impact the clinical workflow in a significant way starting in 1999. The two key catalysts that have propelled increased adoption within healthcare institutions are FCC regulations, as well as the evolution of the IEEE standards, and increasing maturity of the Wi-Fi Alliance. The other two major organizations that have helped push adoption are the Food and Drug Administration (FDA), and the Association for the Advancement of Medical Instrumentation (AAMI). Recent federal government mandates like the push to attain meaningful use have also contributed to driving increased adoption. Many areas have been impacted by mobility, including devices supporting voice and video, but the area that has seen the most dramatic workflow improvements is the medical device arena. With wireless medical telemetry systems (WMTS) on the decline, using Wi-Fi as a means of transporting data from medical devices to the network, and between sensors and medical devices, has been a growing field. Medical device vendors continue to struggle to integrate Wi-Fi into their devices, with hit-and-miss results. Prior to diving into specific use cases, the following section will address the roles that the various government and regulatory agencies have played in shaping the Wi-Fi-centric mHealth arena.

The FDA is heavily involved with clearing different types of medical devices to be introduced to the U.S. market. The Medical Device Amendments Act of 1976 lays the foundation for the 510(k) process, which is used to clear upwards of 90 percent of medical devices to be sold in the U.S. market. Thankfully this process is not as stringent as the processes that are used to introduce a new drug to market. Medical devices are classified into one of three classes as follows.

Class I: Devices that are not intended to sustain life do not require undergoing the 510(k) process or clearance but needing to follow general controls. Tongue depressors and latex gloves are examples of Class I devices.

Class II: Devices that need to meet minimal performance requirements and need to be cleared for safety and efficacy using the 510(k) process. IV Pumps are a Class II device.

Class III: This class of devices is necessary to sustain life, and must undergo the 510(k) premarket approval process, and are often used in clinical trials prior to release. These include devices such as defibrillators and implanted medical devices.

Generally only Class II and Class III devices will require network connectivity and thus can potentially leverage Wi-Fi. The 510(k) process is often lengthy and involves substantial testing which is generally focused around patient safety and the efficacy of a given device. Network communications capabilities are often taken for granted and are an afterthought. Areas like how a device will function in a dense Wi-Fi environment, preferred frequency bands, and supported authentication and encryption schemes are generally farmed out to the manufacturer of the wireless card being used, with little consideration for wireless best practices. The line of demarcation between regulating a device as a medical device and regulating it as a communications device has prompted the FDA to work closely with the FCC when dealing with wireless medical devices. In 2011, the FDA released draft guidance on mobile device applications (Medical Device Data systems rule). The integration between these two organizations is crucial for the success of the mHealth space.

The FCC released the MBAN proposal in 2012 which allocates a dedicated spectrum for body sensors to transmit data in real time. The idea is that these types of sensors will result in a substantial return on investment for healthcare institutions by decreasing the risk of infections and promoting early decisions and better outcomes.

Although the FDA is starting to move in a direction that is helping drive mHealth forward, there is still much lacking. When medical device vendors design a device, it often takes upwards of a year to introduce it to market. In the telecommunications space, the span of a year can see tremendous improvements from the perspective of standards, security, or bandwidth availability. By the time a device makes it to the market, the integrated Wi-Fi capabilities are often outdated. The device can have a lifecycle spanning upwards of 5 years, or longer in some instances. It is crucial for these types of medical devices to have a flexible networking architecture that allows for upgrading drivers and even hardware if needed, with minimal scrutiny from the FDA. If the sole functionality being impacted is Wi-Fi functionality, it would be beneficial to have a series of high-level wireless tests that can be conducted to clear the firmware, or even hardware upgrade path.

We only touch the tip of the iceberg when discussing medical devices. A new type of medical device that integrates with smart phones and tablets is really pushing the traditional boundaries with the FDA. This area, compounded by the explosive growth of healthcare-related mobile applications, has been forcing the organization to rethink and reinvent its review mechanisms.

In June of 2013 the FDA released a draft guidance pertaining to the cybersecurity of medical devices. The target audiences were primarily medical device manufacturers, and the document entitled "Content of premarket submissions for management of cybersecurity in medical devices" calls attention to intentional threats to medical devices. These range from Malware and viruses infecting medical devices to organized penetration and Denial of Service attacks. The ruling urges medical device manufacturers to develop a set of security controls to assure medical devices maintain information confidentiality, integrity, and availability. In part, this means implementing two factor authentication mechanisms including passwords, biometric identifiers, or smartcards in order to restrict the number of individuals capable of interacting with the product.

It can be argued that the FCC is one of the key reasons that wireless technology was able to thrive in healthcare. Since the organization released the ISM band for unlicensed use in 1985, and more recently dedicated a portion of the radio spectrum to WMTS in 2000, it laid the foundation for medical device manufacturers to start to focus on this space. The FCC continues to play a fundamental role in driving mobility in healthcare. The organization's National Broadband Plan released in 2010 along with the ruling allocating 40 MHz of spectrum—2360 to 2400 MHz—for use by medical body area networks

(MBAN) devices in 2012 is a testament to this. They have also been involved in creating some best practices documentation around securing wireless devices. In an effort to remain a leader in the mHealth space, in 2012 the FCC announced that it would be adding a position of Health Care Director to continue to drive innovation in this space. The FCC continues to work with the FDA to ensure that available spectrum is allocated to promote mHealth as much as possible. They have been making every effort to foster innovation.

The AAMI has always been a fundamental player in medical device innovation and design. The organization has been developing standards for medical device design for decades. Wireless medical devices have traditionally been viewed like any other medical device. The typical AAMI audiences are clinical or biomedical engineers who generally deal with the maintenance and repair of medical devices. As medical devices become more dependent on networks and make use of Ethernet and Wi-Fi, the organization has been promoting the need for collaboration between IT and clinical engineering. Many healthcare institutions have taken this mantra to heart, and have shifted their reporting structure so that clinical engineering staff reports to IT leadership. This is an inevitable step given the growth of Wi-Fi-capable medical devices.

By leveraging Wi-Fi, medical device manufacturers have ventured into a shared medium that is outside of their control. When one also considers that many medical devices leverage fairly widespread core operating systems, like Windows, the number of variables that can cause data transmission issues grows. AAMI released the IEC 800001-1 series of standards between 2008 and 2012. These are intended to apply appropriate risk management to IT networks that support medical devices. This is in line with ISO 14971. The standards address safety, system security, and effectiveness, which are generally regarded as necessities for patient well-being. It incorporates best practices for risk management as well as change release management. These are in line with ITIL is the most popular and widely accepted approach to service management. It stands for information technology infrastructure library methodology which is well adopted in the pure IT arena. "Accordinding to the AAMI (Association for the Advancement of Medical Information) IEC 80001-1 it defines responsibilities for parties such as medical device manufacturers, non-medical device manufacturers, the responsible organization, IT-network integrator, and potentially others, engaged in installing, using, configuring, maintaining and decommissioning IT-networks incorporating medical devices." There are four key areas that the standard highlights:

- The three risk components to be managed are safety, effectiveness, and security—and in that order of priority.
- It is ultimately the responsibility of the "responsible organization" (typically, the healthcare provider) for risk management of medical devices interacting with an IT network.
- "Responsible organization" includes health-delivery organizations of all size, such as physician single and group practices, as well as hospitals, clinics, etc.
- For the objective of 80001 to be met, the "responsible organization" will need to work closely with medical device manufacturers and providers of information technology.

The AAMI has paved the way for healthcare IT staff to be able to reach out to medical device manufacturers directly and work on fine tuning the network performance of a given device. Some examples of this are highlighted in the use case section of this chapter. The organization continues to provide best practices for managing wireless medical devices in their publication *Biomedical Instrumentation and Technology*. In addition, the AAMI established the Wireless Strategy Task Force (WSTF) in 2013. The group, comprised of manufacturers, regulators, users of technology, and other interested parties—is developing educational resources and tools and sharing best practices to address wireless challenges in healthcare. Group priorities include clarifying roles and responsibilities in the wireless arena, managing spectrum to improve safety and security, designing wireless infrastructure for high reliability, learning from other industries, managing risk and preventing failure. The group released a special compilation of articles in 2013 entitled "Going Wireless", which is a great resource for anyone working with mobile medical devices (https://www.aami. org/hottopics/wireless/AAMI/Going_Wireless_2013.pdf).

There are many other organizations that can be mentioned in these sections, such as the National Institute of Standards and Technology (NIST), the Healthcare Information and Management Systems

Society (HIMSS) and its mobile initiative mHIMSS, and the federal government, but the last one that will be discussed is the Wi-Fi Alliance. The background of this organization was discussed in the introduction, but for the purposes of this chapter, it is important to note that the Wi-Fi Alliance has been instrumental in publishing guidelines for deploying, securing, and leveraging Wi-Fi in healthcare.

New wireless medical devices are a blessing; they can also be difficult to troubleshoot, as many large medical device manufacturers such as GE, Medtronic, Philips, Baxter, and CareFusion, are designing and adapting medical devices for use on unlicensed radio frequencies. Often, manufacturers will cut costs by using noncompliant or out-of-date wireless devices (adapters, bridges, etc.) embedded in the medical devices. This effort to reduce cost and to gain market share has been a growing challenge for network administrators in healthcare. From diagnostics and monitoring, to the operating theatre and managing patient medical records, demand on wireless technology is more complex and mission critical in the healthcare industry. As medical device manufacturers race to introduce new devices, in many cases they must adhere to HIPAA-HITECH requirements and the FDA's 510(k) approval process. Healthcare organizations often face a lack of central control over procurement because departments have their own budgets and purchasing power. As ubiquitous Wi-Fi is becoming a reality, it is increasingly challenging to manage existing and legacy wireless medical devices while continuing to drive forward and utilize the latest available technology. Often manufacturers will take shortcuts by introducing an add-on Wi-Fi integration using wireless bridges, or will opt to utilize lower-end, cheap wireless cards in their equipment. This makes managing wireless medical devices a challenge requiring a close working relationship between clinical engineering and IT.

When it comes to patient data, securing medical devices and their data is vital to providing safe and effective healthcare. As Wi-Fi is growing the risks associated with the technology are inherent and are becoming more lucrative for hackers to try and take advantage of. Some of these risks are associated with security, availability, quality of service (QoS), and privacy. As the healthcare industry continues to expand and enter the ever-growing wireless space, including patient monitoring equipment, physicians' PDAs and laptops, and wireless-enabled medical devices, the risks associated with their use also rise. Some healthcare organizations have stayed ahead by deploying secured wireless networks for their medical devices. They often have to tweak their network to accommodate nonstandard or legacy medical devices.

Different organizations and departments within the hospital often mandate the wireless medical devices to purchase. In order to avoid a chaotic situation, they must be required to utilize risk management techniques and to thoroughly test each and every device that is being proposed for deployment on the Wi-Fi network. If any of the devices cannot meet minimal security requirements, they need to be identified.

The rapid pace of wireless medical device procurement presents an opportunity to create a focused certification process for the wireless medical devices. The certification process entails thoroughly testing the wireless medical device, and clearly identifying clinical workflow and support expectations. The IT department and clinical staff can work together to create a detailed inventory of all the wireless medical devices deployed in the hospital. Once that is done an OLA (operational level agreement) and SLA (service level agreement) can be set up to describe the maintenance and support matrix for each type of device. Proper planning and design are important to ensuring that the wireless network will support certain devices. Healthcare institutions wishing to manage their wireless medical devices should develop a consistent process for onboarding devices as well as phases for bringing all of their wireless medical devices up to a minimal set of authentication and encryption requirements.

The current industry consensus is that the best practice for wireless medical device authentication and encryption is using 802.1x with EAP TLS and AES encryption. This enforces mutual authentication and requires each medical device to have an x.509 certificate installed before it is allowed onto the wireless network. Due to the wide spectrum of device wireless capabilities, it is often necessary to use a phased approach to manage wireless medical devices and promote ongoing authentication and encryption best practices. HIPAA advisory and wireless interoperability-certifying Wi-Fi Alliance has acknowledged that the typical 802.11 security features such as WEP and/or shared key authentication are not secured enough. The phases are outlined in the bullet points below:

- **Phase 1:** All medical devices that support a certain authentication and encryption should be configured to use a dedicated SSID, keeping the number of SSIDs as low as possible. This phase is targeted at minimizing the amount of wireless overhead traffic. IT and clinical engineering staff need to consolidate a detailed inventory of all wireless medical devices in the hospital. This should include the make and model of the device, network connectivity requirement, device classification, supported spectrum, and high bandwidth requirements. This process will provide more insight into which wireless medical devices are capable of handling and supporting certain authentication and encryption methods.

- **Phase 2:** The purpose of the medical device policies on the network is to ensure that each device is suited for its purpose and meets clinical and patient needs, to make sure that the device complies with safety and quality standards. Since medical devices are regulated by the FDA, their design and operation cannot be modified by the end user. For many years, device manufacturers have been responsible for the installation, service, and support of their devices, including the network. This has resulted in several small independent networks in the hospital. As wireless technology continues to expand, hospitals feel the increasing financial pressure to deploy medical devices on their existing enterprise network. Network policies need to be applied to limit medical device network access to required IP addresses.

- **Phase 3:** Continuously refresh medical devices that do not support WPA2 EAP TLS. This should eventually result in one SSID using EAP TLS.

- **Phase 4:** Implement EAP TLS. The complexity associated with deploying EAP TLS is dependent on whether the hospital has a PKI and a certificate authority in place. Building such a system can be an expensive undertaking.

- **Phase 5:** Develop an overall stringent wireless security policy for medical devices that is interdepartmental and ties into IT governance, security, and procurement. Part of the policy needs to be ongoing device certification as a part of onboarding.

The questions that should be posed when evaluating a new wireless device can be broken down into three test categories, functional, network, and failover/redundancy. In addition to these, a detailed risk assessment of the device should be clearly documented. The bullet points below can serve as a starting framework for each of these categories.

Functional Testing

The first series of tests are intended to validate that the wireless medical device being evaluated is IEEE compliant. The following should be validated as part of the test:

- Is the wireless-capable medical device designed to be mobile or stationary?
- Does the device operate in the unlicensed RF spectrum?
  - Is it IEEE 802.11a/b/g/n or any subset thereof compliant?
  - If not, what RF frequencies does it utilize?
  - Is it Wi-Fi certified?
  - What PHY rates are supported?
  - Is the wireless capability provided by a bolt-on bridge or an integrated wireless card?
  - What models of wireless card and chipset are used?
  - What is the average packet size transmitted, and the maximum latency and jitter requirement?
  - Does the wireless card on the device support "super frames"/frame aggregation (802.11n)?
- Is the device IEEE 802.11i compliant?
  - Is WPA2 encryption supported?
  - Does the device support 802.1X?
  - What types of EAP can the device support?
  - Can the device be added to a Windows domain within Active Directory?
- Is the device IEEE 802.11e compliant?
  - Does the device support WMM and/or WMM PS Mode?
  - What queue is recommended?
- Is the device IEEE 802.11r compliant?
  - Is fast secure roaming supported?
  - Is Opportunistic Key Caching supported?

- Can the device firmware be updated as wireless authentication and encryption mechanisms evolve in the industry?

Network Testing

The tests/questions below are oriented toward understanding the impact of the proposed device on the wireless and the wired networks.

- Does the device support dynamic host configuration protocol (DHCP), or does it require a static IP address?
- What type of information is transmitted via the wireless medium?
  - Is it required for the device to be on the corporate wireless network?
  - What does the device need access to on the corporate network? Can you list all appliances and necessary TCP/UDP ports?
  - Does the device transmit ePHI (electronic protected health information)?
  - What is the network bandwidth requirement for the device?
  - Can the maximum transmission unit (MTU) size be manually modified on the device if needed?

Failover and Redundancy Test

- In the event that there is a disruption to the wireless network, what actions are taken by the device?
  - Does it support or provide a backup mechanism for transmitting data if needed?
  - Does it automatically try to retransmit the data once network connectivity resumes?
  - Does it have a password-protected administration mode for modifying network settings?
- If the device loses network connectivity, will it directly impact a life-sustaining ongoing process or procedure?
- Does the device support removable storage media?
  - Is USB or Firewire supported?

- Does the device have an accessible/removable hard drive?
- Does the device store ePHI on removable media?

The IEC 800001 standard clearly outlines a process for assessing the risk of using a given medical device on the network. Some of the areas that need to be clearly understood are things that can go wrong with the device resulting in unintended consequences. A risk acceptability matrix should be created for each wireless medical device being introduced to the network.

Wireless medical devices have gone through several design iterations, some of which are still around. WMTS and personal area networks are a few of the wireless technologies in use, but Wi-Fi seems to be the one medium that will last. The bandwidth available is significantly higher than any of the other wireless technologies. Cellular providers, who were against leveraging Wi-Fi by dropping their traffic locally onto corporate networks have been feeling the bandwidth crunch and are now fully supportive of Wi-Fi offloading.

In the next section the use cases with various types of medical devices, observations, and lessons learned are based on real-world experiences.

Mobile X-Ray Machines

Mobile x-ray machines are one of the first types of devices that have taken advantage of the wireless network to transmit images to a central server. These devices can be wheeled into patient rooms, and can be used to provide on-demand x-rays at the bedside. This is a significant workflow improvement over the older dedicated room with an x-ray device. In the early 1990s, it would often take a physician several hours to have a patient wheeled to the x-ray room and then have the final x-ray print in hand. Mobile x-ray devices can cut that process down to less than half an hour. These units can often be bulky, and they are fitted with a lead apron, motorized wheels, and often have an Ethernet connection as a backup way to upload x-ray films (Figure 6.1).

From a network traffic perspective, this type of device can be demanding on the wireless network due to the size of the files that need to be transmitted. These can be several megabytes large.

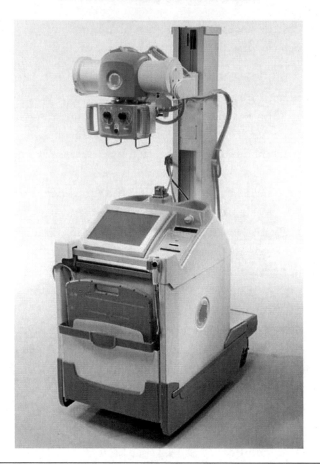

Figure 6.1   Example of a mobile x-ray machine.

Evaluating one such device by a prominent medical device manufacturer led to the following obervations:

1. The device had a built-in wireless card installed into the card rack of the computer in the bottom of the chassis. The signal output is set to its maximum setting and is not adjustable from the software. The wireless antenna is located under the top cover, below the LCD screen, and is vertically polarized.
2. Due to its poor wireless design, the positioning of the cart and its orientation relative to nearby access points significantly impacts its receiver sensitivity. The card has a stronger signal when the LCD screen is facing the nearest access point. The device does not support AES encryption but rather reverts to TKIP.

3. While rolling the mobile x-ray device around various parts of the hospital, it was noted that the roaming aggressiveness of the device is fairly low, which tends to cause the cart to roam very poorly. In effect, the x-ray device roams when the signal from the access point it is associated to becomes unusable, which tends to make the roaming process choppy.
4. The device does not have a wireless survey mechanism to assist IT professionals in determining the RSSI values that the device is detecting.

Medication Dispensing Systems

Medication dispensing units are generally used to securely store and tighten control over medication in hospitals. They also help manage errors associated with delivering the appropriate medication to the right patient. These can be found in various areas throughout the hospital ranging from inpatient floors to the intensive care unit and operating rooms. Generally, these resemble a large chest with drawers, with an imbedded or overlay monitor, with an integrated barcode scanner and sometimes a printer. Some of the newer iterations are smaller and feature Wi-Fi connectivity, but the nature of the device demands adequate storage space.

The reasoning behind using Wi-Fi to provide network connectivity is to avoid having to run a dedicated Ethernet cable for the device. Although this sounds great in theory, one needs to factor in the other devices utilizing the shared wireless medium, and the true necessity for the drug dispensing unit to be mobile.

Evaluating one such device by a prominent medical device manufacturer led to the following observations:

1. Although the device had an integrated wireless card, it required connection to an AC power outlet to function. This requirement limits the mobility of the device, and makes one question whether it truly needs to be on the wireless network.
2. These carts are generally left in one area, and can easily use a wired Ethernet connection, freeing up valuable wireless bandwidth.
3. If there is a strong enough demand for these devices to utilize the wireless network, it is essential to certify the functionality

of the device on a given network. We ran into an issue where one of these devices did not properly support opportunistic key caching, and was running dated wireless drivers.

Due to their typical use case, it is more cost effective in the long run to plan to have wired connectivity for these devices as opposed to having them utilize the Wi-Fi network.

IV Pumps

The invention of the wearable intravenous (IV) infusion pump by Dean Kamen in the 1970s was a major catalyst for medical device engineers to start looking into ways to keep these types of medical devices connected to the network while being mobile. Infusion pumps can be used in scenarios ranging from basic hydrations to blood transfusions, or efficient delivery of medicines.

IV pumps (Figure 6.2) are one of the most heavily deployed wireless medical devices within hospitals and hospital systems. Their small form factor and modular design make these ideal for mobility. The Wi-Fi capability built into these types of devices ranges based on how well the unit was designed to be mobile. Some IV pumps that were released in the early 2000s are unable to support WPA2 (AES

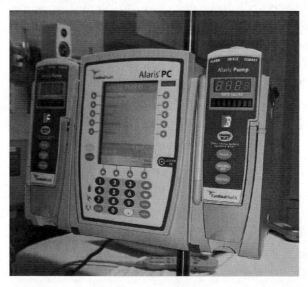

Figure 6.2   IV pump.

encryption), but newer models can support most forms of authentication and encryption. The network is utilized as a means of collecting and trending data from a given pump as well as a way to keep drug libraries up to date on a given pump. When dealing with multicampus or global wireless network deployments that rely on centralized controller architecture, it is important to understand the data set dependencies per site. For example, even if two different hospitals have access points that are hosted on a given controller, it is sometimes necessary to provision unique VLANs for each to ensure that the pumps at each hospital receive their intended drug library/data set.

The wireless hospital system where the following observations were made is a fairly large system, which had approximately 2000 wireless IV pumps. These were distributed among the various hospitals and ranged from 900 at one site down to a handful at some of the smaller clinics. The pumps relied on a built-in wireless card which behaved very similarly to a Wi-Fi card on a given laptop. Earlier releases of code had issues with WPA2 encryption, but this was quickly corrected with a firmware update. The pumps have worked very smoothly and are very conservative in their bandwidth utilization. Furthermore, the wireless connectivity does not impact the core fluid pumping functionality of these devices. One of the key observations when dealing with these types of pumps is the necessity for a detailed inventory, ensuring that they are under maintenance and are running the latest firmware, and that the wireless network is able to provide high-level location tracking for these devices.

The largest hurdle when dealing with wireless IV pumps is revisiting the wireless network architecture to ensure that it can accommodate the drug library/data set push to each pump.

One topic that was touched upon in an earlier chapter is Real-Time Location Services (RTLS). This plays a significant role in the IV pump inventory and workflow management. It makes it easier to ensure that each IV pump is cleaned between uses rather than being moved from one room or floor to the next without any formal reconditioning and cleaning. Many hospital systems lose track of the locations of their IV pumps and start renting these devices at an astronomical monthly cost. With a finely tuned RTLS system, these recurring expenses can be eliminated.

Electrocardiogram Carts

Electrocardiogram (ECG) carts are often part of a larger system managed by the cardiology department. These can have a mobile form factor that can be rolled into a patient room and used for cardiac tests at the bedside. These are typically comprised of a mobile cart and several cardiac leads (Figure 6.3).

The model discussed here is manufactured by one of the largest medical device companies in the world, so some of the observations are quite alarming. The ECG device we were asked to bring onto the wireless network had some unique requirements which are puzzling to this day. These units do not have a built-in wireless card, but rather rely on a bolt-on wireless bridge. The bridge is a stand-alone device which is configured independently of the ECG device. It connects to the Ethernet port on the device with the intent of tricking the system into believing that it is connected to a wired network jack. The unique

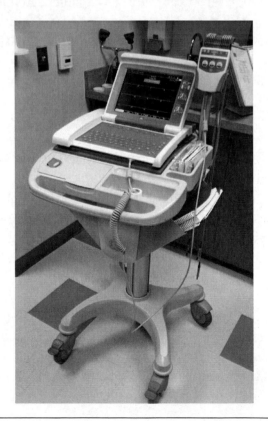

Figure 6.3 ECG cart.

requirements were for a reserved DHCP IP address, or a static IP address per device. It was also brought to our attention that the device does not support AES encryption, but rather supports TKIP.

The lack of support for AES can be addressed with a firmware upgrade, but the requirement that stood out was the IP addressing requirement. Any engineer who has worked in a large-scale environment understands that different intermediate distribution frame (IDF) closets on different floors of a facility can sometimes require different VLANS for capacity planning. This means that a device with an IP address on one floor cannot seamlessly roam to a different floor unless the same VLAN spans both floors. The wireless space is no different. If a client receives an IP address from an access point on one controller, it cannot seamlessly roam to an AP on a different controller without creative network architecture. In the case of the ECG devices, this means that they can only function in certain geographic areas and cannot roam throughout the hospital. The word "static" IP address should throw up a red flag when one is working on mobility.

Ultrasound Devices

Mobile ultrasound devices come in various shapes and sizes depending on their intended use. These range from a handheld tablet to dedicated workstations on wheels (Figure 6.4). The form factor is in part dependent on the function of the device and its required signal-processing capability. These devices provide clinicians with the ability to view subcutaneous activities ranging from potentially damaged organs to cardiac issues, and viewing the fetus in expectant mothers. All of these units rely on introducing high-frequency acoustic energy, and analyzing the return signals to generate images. They generally rely on dedicated transducers to analyze and digitize the return signals. The higher resolution units are generally not battery friendly, hence the onboard battery units. Portable units sacrifice performance for a longer battery life. From a mobility standpoint, both types can be integrated onto Wi-Fi networks. The size of the images and videos captured can range from several megabytes for high-resolution images to a dozen or more megabytes for videos. Historically these units have featured removable memory media for storing images, which can be used to transfer images. With growing

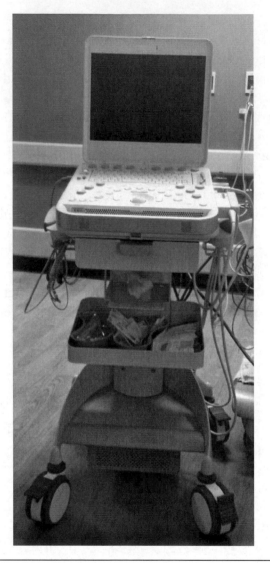

Figure 6.4   Mobile ultrasound machine.

patient privacy concerns, the newer devices have removed this capability, and only allow buffering capability in the event that network connectivity is lost. These devices rely on a central storage server for categorizing and storing images.

Different departments in the hospital deploy a preferred form factor based on their needs. A cardiac or a prenatal ultrasound device is generally integrated into a larger mobile cart with several dedicated transducers. The ER department hosts a variety of these to meet the various needs. The larger units have the same limitation as other mobile medical devices. They rely on wireless bridges or USB wireless cards. These pose a challenge for seamless roaming. The devices manufactured in the last 3 to 5 years host integrated wireless cards. The combination of old and new devices imposes a significant support burden on IT and clinical engineering. The tablet form factor devices integrate more readily and are easier to support with their integrated IEEE 802.11 a/b/g/n and 802.11i support. These can be used as a first line of diagnosis, but to avoid liabilities the testing is supplemented with the higher end units capable of more advanced signal processing.

A variety of handheld ultrasound devices have been introduced since 2010. These will be discussed in more depth later in the chapter, but many hospitals are starting to rely on these in place of the traditional stethoscope. A growing number of physicians are relying on these devices at the bedside to augment the ability of the traditional stethoscope. They have proven to be more effective for diagnosing conditions like pneumonia. With the proliferation of smart phones and tablets, these devices are beginning to rely on these form factors. There is still much work and research to be done in order to standardize these devices.

Blood Gas Analyzers

Blood gas analyzers range from benchtop to portable units, but they all perform the function of measuring base status, ventilation, and arterial oxygenation. The unit in this use case was designed to be mobile on a dedicated cart (Figure 6.5).

It did not have a wireless card and we managed to work with the vendor to migrate the device from relying on a wireless bridge to having a nano-USB based 802.11 b/g wireless card. Ultimately, the device will need to be transitioned to the 802.11a network to free up capacity in the 2.4-GHz space.

Hemodialysis Machines

In the hemodialysis machine space, a growing trend is to leverage the same device for patient entertainment as they undergo a procedure

Figure 6.5   Blood gas analyzer.

that can take several hours. Devices that are prevalent in dialysis centers are about the size of an ATM machine with an onboard PC (Figure 6.6). Patients can use the PC portion of the device to access the Internet.

More than 400,000 Americans receive dialysis, about half of them over age 65. More than 90 percent go to dialysis centers for professional care, but the home dialysis options are beginning to take root, which is prompting some interesting, portable form factors. These units are fairly invasive, and require fine-tuning, so many patients shy away from them.

PC and dialysis functionality are logically separated, with the PC relying on Wi-Fi while the hemodialysis device leverages a serial interface. Data is not correlated to a specific patient until it reaches a central data repository. The Wi-Fi capability on these machines is required for the onboard PC, which typically runs a standard operating system like Windows XP, or Windows 7 with an integrated

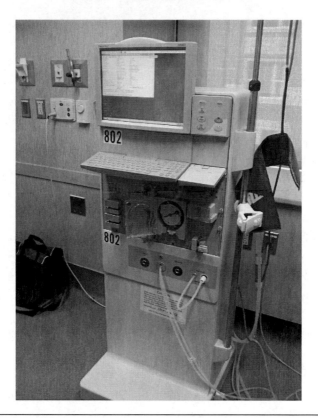

Figure 6.6 Hemodialysis machine.

wireless card. The card is fairly flexible from a configuration stand-point, and tends to roam as well as wireless cards that are found in a typical laptop device.

mHealth

With the growing focus on the empowered patient, many of the devices covered in this chapter are becoming increasingly mobile. Dr. Erik Topol, a pioneer in this space, demonstrated the power of the technology during a keynote address at the HIMSS conference in 2013 in New Orleans. Ironically, he put the AliveCor Heart Monitor, one of his showpieces, to use on a fellow passenger in distress shortly after his talk. When we combine the capabilities of some of these devices to help patients become more aware of their health with the growing trend of patients being able to review their medical records via cloud-enabled portals, it becomes clear where the technology is headed. The challenge for medical device manufacturers is to leverage authentication and accounting as well as provide a secure means of transferring this type of data to physicians. This trend has the potential to change patient care as we know it. Interestingly enough, a recent survey by Harris Interactive revealed that doctors are not supportive of full transparency. A survey of 3,700 doctors in eight countries showed that only 31 percent believe that patients should have full access to their own medical records via electronic means. The majority of those surveyed, some 65 percent, supported restrictive access. One can empathize with these results given that the vast majority of patients are not qualified to fully understand the results. What may appear to the patient as a major issue can sometimes be an anomaly, and only a trained physician can make the distinction. If one compares this to an adage in IT, this can be likened to someone reading a packet capture and analyzing it, and arriving at a root cause of a network issue.

With the leaps in artificial intelligence and the computational capabilities in line with Moore's law are enough to make one take a step back and ponder the future of patient care. After all, Watson, the IBM supercomputer has taken up healthcare. With the ballooning, unsustainable costs of healthcare, it may be these types of technologies that can help remold this field and make it sustainable. There is no denying that the per capita spending has little correlation to increased life expectancy. With a heavy focus on wellness, it is conceivable that mHealth applications will play a major role in allowing patients to assist in regulating their own health by becoming increasingly aware of their health.

The challenge to IT as portable mobile medical devices become more prevalent is to ensure that all relevant security guidelines are followed and to prioritize traffic appropriately based on applications rather than device types. If this is not managed properly, there could be a negative impact on patient care.

One area that is disturbing is our growing dependence on smart phones and tablets, and an augmented reality approach with a dependence on a smaller form factor will be a breath of fresh air.

7

VOICE OVER WI-FI

Voice over Wi-Fi is a challenge to implement and manage in any environment, but there are many factors unique to healthcare that present even greater challenges. Voice over IP or VoIP was birthed in the idea of reducing the cost of a telephone call. Voice over Wi-Fi or VoWi-Fi similarly aims to reduce mobile phone costs. This is no surprise as the entire world seems to be drawn to mobile communications. Some companies and organizations have foregone installing traditional desk phones and solely utilize cellular communications. Designing and implementing a VoIP network that is reliable and scalable is very difficult and the challenges are multiplied when the transmission medium is 802.11 based. With the complexities in design and support, this may not end up being a large cost savings as one might expect.

Why VoWi-Fi?

This may seem like an odd question to ask in a book entitled *Wi-Fi Enabled Healthcare*. Perhaps it is human nature to explore the limits of nearly everything around us. Ever since Dr. Bell called for his assistant, Mr. Watson on that epic day the telephone came to fruition, we have been improving communications technology. The evolution of the technology and the increasing use cases seem to have no end. That said there are alternatives to VoWi-Fi. An older solution that is still commercially available uses the ISM 900-MHz band of the spectrum. Another alternative is something called digital enhanced cordless communications (DECT). This is a technology that is built for in-building voice communications. It has been much more widely adopted in Europe than in the United States. This solution may sound familiar as many people have cordless home phones that use it. Commercially, it is branded DECT 6.0 which causes some to think this uses 6.0-GHZ communications, but it actually uses the

1.9-GHz spectrum. The spectrum each of these technologies use is referenced because it is important for comparison purposes. Not all of the RF spectrum is created equal. Lower frequencies are attenuated less than higher frequencies at the same power level (in most cases). Sometimes it is said that lower frequencies travel farther. While this is not exactly correct, the observed effect is that lower frequencies can be used over longer distances and generally are. There are many factors that affect the distance of wireless communications. Usually, the lower a frequency, the less it is attenuated and therefore the further it is able to travel. The RF spectrum can be likened to real estate and much of the lower frequency range would be the spectral equivalent to the city of Manhattan. There is a finite amount of it. Spectral efficiency may not be a common term for most people but it is important to wireless engineers. Take a typical 802.11n 40-MHz wide channel for example. If this transmits at a data rate of 300 Mbps and uses 40 MHz of spectrum, the spectral efficiency is 7.5 bits per second per hertz. This is the limitation of lower frequencies. There is less spectrum available and it is incredibly valuable due to its attenuation properties. Low-frequency communications are expensive to use in higher throughput applications like video. This is why VoWi-Fi matters. No other spectrum is capable of providing both voice and high throughput data communications in a spectrum that is unlicensed. If a lower and more plentiful spectrum existed, we would be writing about that instead. While it is not perfect and certainly not designed from the start with voice communications in mind, VoWi-Fi will be around for the foreseeable future.

The Challenges of VoWi-Fi

There are three main categories that the challenges with designing, managing, and troubleshooting VoWi-Fi can fall into: radio frequency physics, quality of service, and scalability. Wi-Fi uses a technology called spread spectrum. With spread spectrum the entire frequency range is broken up into separate channels and transmissions are spread across a channel. This makes it possible for multiple devices to transmit and receive at the same time without interfering with each other. 802.11 b/g/n uses three nonoverlapping channels (in the United States) and 802.11 a/n uses as many as 24 channels. The

2.4-GHz spectrum has too few channels and the 5-GHz spectrum has too many for voice to work ideally. Let's start with the 2.4-GHz spectrum. With only three channels available, adjacent access point (AP) interference can be an issue. Many best practice guides will indicate that a minimum Relative Signal Strength Indicator (RSSI) of −67 dBm is needed for a VoWi-Fi network. This requirement comes from ensuring that only the maximum data rates are used. If two APs on the same channel are too close or their power levels are too high, the potential for interference increases. In order to avoid low data rates and retransmissions it is best to ensure that interference is all or almost all eliminated. Retransmissions are far worse than low data rates. A retransmission can occupy the medium for very long periods of time. Wi-Fi uses modulation, which is a way of encoding wireless transmissions to be resistant to corruption during transmission. Each modulation type and encoding combination requires a certain ratio of modulated signal to noise. Noise is something that exists everywhere. The ambient RF noise is at the same level at every part of the spectrum. For example, in order to maintain a 54-Mbps data rate with 802.11g, it is necessary to have a 24-dB signal-to-noise ratio. If the noise floor is −91 dBm this would require the minimum RSSI to be −67 dBm. However, if an adjacent AP can be heard at −85 dBm the minimum RSSI would actually need to be −61 dBm. Some specifications cite even tighter requirements, as high as 30 to 35 dB signal-to-noise ratio for higher data rates. Data rate selection algorithms play a big role in voice over IP. Many engineers believe that Wi-Fi radios mostly use radio measurements to determine what data rate to use. This may be the case for the noise floor but random interference from sources other than adjacent APs could be hard to predict. As a result the more likely action is to use retransmission as the data rate selection method.

The minimum Wi-Fi packet exchange needed to transmit data successfully is called a two-way handshake. After a client or AP has gained access to transmit on the medium, it will send a wireless data frame. Upon reception at the intended recipient and passing an integrity check, the recipient will transmit an ACK frame or acknowledgment. The ACK notifies the first device that the data rate was good. If an ACK is not received, typically the client will retransmit at the same data rate. If an ACK is again not received then the data rate will begin decrementing. The only way to know for sure that voice traffic will be able to traverse a wireless LAN without corruption and retransmission is to validate a survey based on actual data rates from the VoWi-Fi handsets to be used.

Planning the 2.4-GHz spectrum for VoWi-Fi is challenging. The AP placement and power needs to be just right. In many cases it may not be possible to get it perfect everywhere. Sometimes proper AP placement is not possible and compromises have to be made. If too many compromises must be made in the 2.4-GHz spectrum it may be a good idea to use the 5-GHz spectrum for the VoWi-Fi deployment.

The 5-GHz spectrum is better suited for voice traffic. This is primarily due to the additional spectrum that is available. Wi-Fi networks have up to 255 MHz of spectrum available for use. An 802.11a channel has a 20-MHz bandwidth, which provides 24 nonoverlapping channels. Having this many channels to choose from makes channel planning and noise separation fairly easy. The Achilles heel to using 802.11a is wireless roaming, which has been quite a problem for VoWi-Fi. It seems that when a VoWi-Fi phone stops sending or receiving voice frames for more than 50 ms, doctors and nurses tend to call their IT support and report poor call quality or even dropped calls. 50 ms does not leave a lot of room for error. In terms of channel selection in the 5-GHz spectrum once again all spectrum is not created equal. It turns out that even within this free-of-charge spectrum we can find certain channels better suited for voice applications than others. First, the FCC created special rules outlining how the UNII-2 and UNII-2 extended bands can be used. Figure 7.1 shows all the 5-GHz channels available in the United States.

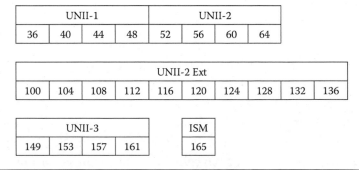

UNII-1				UNII-2			
36	40	44	48	52	56	60	64

UNII-2 Ext									
100	104	108	112	116	120	124	128	132	136

UNII-3				ISM
149	153	157	161	165

Figure 7.1   The UNII table.

These channels are subject to DFS or dynamic frequency selection. DFS bands require any equipment operating in them to scan for weather radar operation. If a radar blast is detected the device must leave that channel for a minimum of 30 minutes. As you may imagine, this can be very disruptive to voice communications. That is not the only reason to reduce channel-planning usage to the eight channels available to the UNII-1 and UNII-3 bands. The roaming process begins with 802.11 beacon frames. Every client has an algorithm that decides when the client needs to find a new AP to associate to. There are a number of factors such as data rate, RSSI, and retransmission counters that can help a client device determine if it needs to find a new AP. When the time comes for the client to begin searching, it will likely use two primary tools: beacons and probes. Beacons are used to notify clients about the wireless network's capabilities. They are transmitted every 102.4 ms or every 100 TUs or time units. If there are too many APs available to the client this could prevent a client from making the correct decision for a roam. A client could have to listen on a channel up to 102.4 ms to hear a beacon from a nearby AP. While 100 ms is just a blink of an eye, as mentioned earlier there is only a 50 ms error margin for roams. For this reason client devices typically scan for beacons all the time and keep a list of the nearest APs. Another wireless frame used for finding wireless networks by a given client is called a probe. There are two types of probes: probe request and probe response. Essentially, these are broadcast out to any APs that are listening one channel at a time. The client will begin on one channel, send the probe request, and wait for the probe response. With the 5-GHz spectrum, it may take a long time to probe on all 24 channels. This presents a Goldilocks scenario where too much or too little channel use can cause problems. Selecting the UNII-1 and UNII-3 bands for a voice deployment is a general best practice for most voice vendors. There are four 20-MHz channels in each of these bands. Eight channels provide plenty of room for channel planning and there will likely only be two or three of these channels audible to the client. The trade-off for this simplification of the wireless LAN is that it limits the option of using 40-MHz wide channels from the 802.11n standard and allowing for 300 to 450 Mbps data rates in current network equipment. All clients are limited to 20-MHz channels, which cuts the available throughput of the network in half.

There are some further physics and regulatory limitations of this spectrum. Each UNII band has different limitations imposed by the FCC. The UNII-1 band for example is limited to 50-mW output or 13 dBm. This limitation in combination with the attenuation characteristics of this spectrum creates a relatively small RF footprint. Most WLAN and VoWi-Fi vendors recommend all APs be at the same power level. Due to these limitations, a higher density of access points is required for 5-GHz voice designs. These factors should greatly impact the purchasing decision of any organization wishing to improve workflows, reduce cost, and leverage the latest in technology. Voice over Wi-Fi should not be considered as a communication standard in an organization unless the WLAN has been designed and optimized for voice.

Quality of Service Fundamentals

Quality of Service (QoS) is an absolute necessity for facilitating voice packets over Ethernet and Wi-Fi networks. It would be fantastic if there was a magic QoS button on every piece of network equipment that just made everything work perfectly together, but the reality is that Ethernet and Wi-Fi networks were developed without much consideration to running voice traffic over them. Voice over IP and voice over Wi-Fi are usually afterthoughts. As a result, there are a number of standards and very carefully thought out methods of delivering voice over data connections. It can be difficult to make sense of them. In order to provide an understanding of how to provide voice over Wi-Fi, it is necessary to discuss the ins and outs of wired QoS. This will hopefully provide a minimum level of understanding of VoWi-Fi.

Evolution of QoS

As if QoS alone was not complicated enough, there are several organizations that create standards for network communications. The IEEE is a well-known standards body but the IETF (Internet Engineering Task Force) also creates standards through requests for comments (RFCs). The RFC has two purposes. The first is to allow new ideas to be fully vetted by a number of experts in networking. Second, it allows standards to be established and published. As a result, there tends to be an evolution of RFCs. Often, there will be a line in an RFC stating "replaced by RFC1234" or "replaces RFC1234." The reason that I mention this is because there have been a number of RFCs that have addressed the problem of providing determinism in IP and Ethernet communications. Anyone who has ever tried to read the standards from the RFCs would have found quite a few of these that have spanned many years. Here are a few of the IETF RFCs that address QoS.

- RFC791
- RFC1122
- RFC1349
- RFC1455
- RFC2474
- RFC3168

IEEE 802.1Q used bits on the Ethernet header to create a low granularity of determinism on a network. This approach worked fine but came with some scalability hurdles. New RFCs began to address ways to overcome the scalability concerns of using Ethernet headers. One of the first is RFC791 which defined a one byte field in the IP header using 3 bits for IP precedence and 3 bits for Type of Services (ToS). RFC2474 defined the Differentiated Service Code Points (DSCP).

The Journey of a Voice Packet

Dr. Smith picks up his VoWi-Fi phone, which is now on the third ring. He pushes the talk button, presses the speaker to his ear, and responds, "This is Dr. Smith." After listening a few seconds he responds, "I'm on my way there right now," and then hangs up. A quick call and calm voice do not do justice to the chaos that is ensuing behind the scenes of any VoWi-Fi communication. Not to mention the unbelievable amount of technology employing complex math equations necessary to make a doctor's voice traverse a data network and sound just like it did leaving the doctor's mouth. The human voice creates variations in air pressure that are "heard" using a microphone and converted to electric voltage. A miracle of the modern world called the ADC or analog-to-digital converter creates a series of ones and zeros to represent what was received at the microphone. When a handset creates

Layer#	Layer Name	Example
Layer 7	Application	HTTP, FTP
Layer 6	Presentation	ASCII
Layer 5	Session	Remote Procedure Call
Layer 4	Transport	TCP Port
Layer 3	Network	IP Address
Layer 2	Data Link	MAC address
Layer 1	Physical	Wi–Fi RF Signal

Figure 7.2   The OSI model.

these ones and zeros, which are a digital approximation of the actual sounds being emitted from our mouths, they are passed down to a piece of software called the network stack. Something called the OSI (open systems interconnection) model is a standard used to describe networks and break them down into their atomic components. There are seven layers to the OSI model (Figure 7.2). Wi-Fi only uses two of these layers: the physical and data link. Ethernet resides in the data link layer, IP addresses in the network, TCP and UDP headers in the transport layer, and SIP and RTP headers in the application layer. Wi-Fi and Ethernet are considered to work at layer two, IP at layer three, and TCP and UDP at layer four. Layers one to four are considered the lower layers.

A QoS design must look at layers two and three. Additionally, Ethernet and Wi-Fi use different layer two QoS markings. Interoperability between all of this is possible but not always straightforward. We can begin by walking through the narrative of Dr. Smith but this time from the perspective of the phone and network. Visualize a basic network. It has two APs, two switches, and one router, as well as two VoWi-Fi phones.

What Happens at Phone One   Phone one enters into a QoS Basic Service Set (QBSS) with the AP. This basically means the phone will be using QoS. The operation of a QBSS is discussed later in this chapter. In the 802.11 frame format exists a field called the QoS control field. The 802.11e/WMM standard defines eight priorities that map to four different queues. A chart mapping the possible values to their 802.1D user priority is shown in Figure 7.3.

Priority	UP (Same as 802. 1D user priority)	802.1D Designation	AC	Designation (informative)
Lowest	1	BK	AC_BK	Background
	2	—	AC_BK	Background
	0	BE	AC_BE	Best Effort
	3	EE	AC_BE	Best Effort
	4	CL	AC_VI	ideo
	5	VI	AC_VI	ideo
	6	VO	AC_VO	Voice
Highest	7	NC	AC_VO	Voice

Figure 7.3   UP to 802.11e.

To understand how phone one will correctly apply 802.11 QoS, we have to look at the order of operations prior to transmission. All packets begin at layer seven of the OSI model. They are successively handed down to the next lower layer until the physical layer finally transmits the frame. With each layer, additional information is added to the frame. Networks are typically concerned only with layers one through four. WMM/802.11e QoS priorities map to Ethernet CoS Class of Service or DSCP markings in the IP header. The transmission begins at layer seven with the information that represents Dr. Smith's voice. One unfortunate shortcoming of the OSI model is seen in this example. The model was intended to describe networks and create order of operations. As a result not all layers are transparent. You will see that layer seven quickly turns into layer four before you know it. For this example we will skip layers six and five, which brings us to layer four. Layer four is where protocols like UDP and TCP live. Most voice traffic will be using UDP. TCP requires the acknowledgment of packets prior to additional transmissions. This is great for ensuring that all the data arrives; however, for voice communications this is not entirely necessary since the delay created by the retransmission will create an audible error. UDP, which just transmits packets without verifying reception, is well suited to voice. The UDP header adds additional information such as source and destination ports. These are often used by firewall rules to identify traffic. Port 1-1024 is reserved for specific application like DNS, DHCP, and SIP voice traffic. The fourth layer information can also be used in some cases to provide prioritization at the AP, switch, or router. The next layer to be added is the network layer. This is most commonly occupied by the Internet Protocol (IP). The IP header holds a field called the Type of Service or ToS. This pertains to VoWi-Fi as it is the home for differentiated services or DiffServ markings. Diffserv markings are used by the wired and wireless infrastructure to prioritize the forwarding of packets. Most VoWi-Fi phones will allow this to be configured through a management interface. However, many smart phones will not have the same granularity. For this example we will say that the phone is marking its voice packets with DSCP 46 or Expedited Forwarding. This is a common marking for voice priority packets. After adding the layer three header, the packet is handed to the data link layer. The second layer adds the Wi-Fi header, which includes the QoS control field. The QoS control field has seven possible values ranging from 0 to 7. These values map to Ethernet CoS values as shown in Figure 7.3. The physical layer now translates the VoIP frame into Wi-Fi modulated signal out its antenna to be received by the AP.

What Happens at the Access Point   Once the AP receives the wireless frame from phone one, it strips the Wi-Fi header, making note of the QoS control field, and replaces it with an Ethernet header. The Wi-Fi QoS control field only exists when the AP and client device have WMM or 802.11e enabled. In order for the Wi-Fi QoS priority to be preserved, the AP must decide what OSI layer to translate the marking to. The AP can mark at layer two, layer three, or both. Also, the AP can make this decision based on the Wi-Fi QoS control field markings or the DSCP markings. This interaction is called a QoS policy. This is the chaos of QoS. There are no hard rules for how wireless to wired QoS policy must function. Typically, this has been left up to the network engineers to figure out what fits best for the applications and environment. If QoS is enabled without changing the default settings, this will certainly lead to undesirable results. Not every vendor applies and translates tags the same way by default. The Ethernet protocol has a field that is port of the 802.1Q tag which is used for VLAN assignment. This tag is called the CoS, class of service, 802.1p tag. Either way, the Ethernet frame must have the optional VLAN tag enabled to support layer two QoS. As most APs are only layer two devices, its work is done and it forwards the packet.

What Happens at Switch One　　At this point the voice packet has left the AP and is now entering the Ethernet switch. 802.1Q tagging must be enabled for CoS markings to be trusted by the switch. Most managed Ethernet switches have the ability to trust the CoS markings. By trusting the CoS markings the switch will not alter the packet between ingress and egress. If 802.1Q tagging is not enabled on the switch, the default behavior will be to strip the tagging. Additionally, the switch may be configured to set the layer two CoS markings based on the layer three DSCP markings. Other information such as the TCP or UDP port number can often be used to classify the voice traffic. Dynamically altering the QoS parameters in this manner is sometimes called a policy map. Policy maps can be intricate and difficult to support in scale.

What Happens at the Router　　When a packet is routed the Ethernet header is destroyed and a new header is created for a new subnet. However, when a packet is switched the Ethernet header is preserved. Whether the packet is routed or switched the DSCP markings are preserved because they exist in a higher layer. The ToS field, where the DSCP marking lives, is always present in the IP header. The router will not alter the DSCP markings unless explicitly programmed to. The router needs a QoS policy to prioritize based on the CoS or DSCP tags. The remainder of the journey for the VoIP packet to reach phone two is similar to the steps made thus far but in reverse.

Differentiated Services

Differentiated Services, DiffServ or DSCP, have evolved over the years to provide QoS for IP packet payloads. The goal is to provide a granular, scalable solution that offers reverse compatibility to previous standards like IP Precedence. IP Precedence was an early design for QoS that only used the first three bits of the ToS field. While DSCP allows for 64 different values, IP Precedence provides just eight priorities. Figure 7.4 shows all the possible DiffServ markings.

The figure shows three distinct markings: Assured Forwarding (AF), Expedited Forwarding (EF), and Class Selector (CS). The DiffServ tag lives in the ToS field. This field is 1 byte in length. The first six bits represent the DSCP and the last two are the ECN

PHB	DSCP	TOS Field	CS (PHB)	Drop Prec
Default	0	000000	000 (0)	000 (0)
AF11	10	001010	001 (1)	010 (2)
AF12	12	001100	001 (1)	100 (4)
AF13	14	001110	001 (1)	110 (6)
AF21	18	010010	010 (2)	010 (2)
AF22	20	010100	010 (2)	100 (4)
AF23	22	010110	010 (2)	110 (6)
AF31	26	011010	011 (3)	010 (2)
AF32	28	011100	011 (3)	100 (4)
AF33	30	011110	011 (3)	110 (6)
AF41	34	100010	100 (4)	010 (2)
AF42	36	100100	100 (4)	100 (4)
AF43	38	100110	100 (4)	110 (6)
EF	46	101110	101 (5)	110 (6)
CS0	0	000000	000 (0)	000 (0)
CS1	8	001000	001 (1)	000 (0)
CS2	16	010000	010 (2)	000 (0)
CS3	24	011000	011 (3)	000 (0)
CS4	32	100000	100 (4)	000 (0)
CS5	40	101000	101 (5)	000 (0)
CS6	48	110000	110 (6)	000 (0)
CS7	56	111000	111 (7)	000 (0)

Figure 7.4   Chart showing DiffServ markings.

or explicit congestion notification. The DSCP tag is split into two parts: the per hop behavior and drop precedence. As each part has 3 bits, they both have possible values of 0 to 7 hence, the 64 possible DSCP values of 0 to 63. Notice that all of the values are even as the odd ones are omitted from the standard. If this is a new concept to you, it should seem a little confusing. Because IP Precedence is the senior interpretation of the ToS byte, DSCP needed to provide reverse compatibility. With DSCP, the ToS field is now called the DS field or Differentiated Services field. It has 6 bits to define 64 different values but still provide the proper semantics to the 3 bit IP Precedence field. This was accomplished by splitting the 6 bits DS field into two fields of 3 bits, which are called the CS Class Selector and Drop Precedence, respectively. The first 3 bits, which are the CS or Per Hop behavior, map directly to the IP Precedence values or 0 to 7. For example, a voice packet will likely receive a DSCP value of 46, which is 101110 in binary. The first 3 bits could be interpreted as IP Precedence 5 or 101 in binary. I should note that 6 and 7 were only used for network control traffic and were not configurable. So, an IP Precedence value of 5 would be the highest configurable priority and would be consistent with the proper designation for voice traffic. Now we can look at the inverse of this scenario and see if it holds true as well. If a device that is configured to support IP Precedence only transmits a packet onto the network, what would the DSCP value be? It would look like this: 101000 to the computers, or 40 to us. DSCP 40 has the alternate name of CS5. A class selector (CS#) is the DSCP interpretation of an IP Precedence value. So, if you see these values on the network it is safe to say that a device is likely using IP Precedence. DSCP is designed to provide additional granularity over IP Precedence. This extra granularity is called the Drop Precedence. In most networks, the easiest way to ensure that VoIP continues to work as expected is to add more bandwidth. If adding bandwidth is not an option then a network engineer must decide which packets get dropped and which do not. Hopefully, you are getting the idea behind the drop precedence value by now. Drop precedence is only used on Assured Forwarding (AF), also known as classes 1 to 4. The higher the value the more likely it is to be dropped. VoIP traffic will typically be marked as EF (DSCP 46) or CS5 (DSCP 40). Many smart phone apps mark the voice traffic incorrectly as CS7 (DSCP 56) or CS6 (DSCP 48). It is important to understand how the IP ToS/DS field is affected every step of the way through a network. In addition it is equally important to understand how the packets are treated or "queued" at each of those steps. This is defined as a policy map and usually involves selecting queuing methods.

DSCP values are a crucial part of a QoS design, but the queuing mechanisms actually do the heavy lifting of deciding what packets to buffer or drop. It may be the assumption of a network engineer that switches and routers out of the box are configured to properly handle QoS. Mostly likely they will not strip these tags. However, they will usually implement a first-in first-out (FIFO) queuing scheme by default. Just as the name implies, it does not prioritize any packet over another and simply forwards on packets in the order that they are received. Other queuing mechanisms include Weighted Fair Queuing (WFQ), Priority Queuing, and Weighted Random Early Detection. Switches and routers use buffers to prioritize traffic. If no priority exists, then all the traffic will go to the default queue and follow a FIFO operation. If multiple priorities are specified each priority is stored in a different buffer. Queuing algorithms are designed to determine when packets are transmitted from each queue. If there is congestion, a low-priority queue may fill up because it does not get as many opportunities to empty itself. When this occurs new packets cannot be added to the queue and are subsequently dropped. While dropping packets is generally a bad thing, some situations are worse than others. For example, if a packet is dropped from an FTP (File Transfer Protocol) download, it will simply be retransmitted. This will delay the download by a second or more but overall it is an insignificant error to the human observer. On the other hand, dropped packets could have much more adverse effects on real-time communications like voice and video from. A dropped voice packet could result in a person having to repeat himself, or possibly having the call drop altogether, forcing him to redial. For this reason it is important that the queuing mechanisms and QoS policies be in line with the business application needs. Furthermore, consulting wireless AP, switch, and router manufacturer documentation is a key step in the process of building out a QoS strategy.

802.1Q

Differentiated Services is a scalable, granular solution to providing QoS over Ethernet and Wi-Fi, but it's not for everyone. Some organizations may not require the complexity of providing that granular level of service to individual devices. Much of this chapter has made the assumption that most organizations have the need or desire to provide wireless to multiservice devices such as smart phones. There are many purpose-built devices that only transmit a few types of packets. One example is a Wi-Fi handset. For several years it has been a best practice to segregate the VoIP traffic away from the other traffic. This minimally entails providing a VoIP SSID and one or more VLANs dedicated to VoIP traffic. This approach is perfectly fine for dedicated voice devices. Smart phones seem to have disrupted of this architecture. But with dedicated devices this will work. In this situation, each AP will provide a dedicated voice SSID and only Wi-Fi phones will be configured to connect to this SSID. Every enterprise Wi-Fi vendor supports some sort of mechanism to set all the traffic in an SSID to the same WMM access category. This only affects the downlink traffic which is from the AP to the phones. All Wi-Fi phone vendors support a similar mechanism as well. Now, all communications between the AP and phone will be able to use WMM Voice (VO) queues instead of Best Effort (BE). The big caveat is that all of the traffic on that SSID will receive priority queuing. The phones receive higher priority on their traffic because they are transmitting small wireless frames that must be received frequently and consistently. If non-voice devices are added to the voice SSID, the APs cannot ensure the voice packets will be delivered on time. The voice queue is designed to handle voice packets which are different from say a web video or EMR applications packets. Sure they are made out of 1s and 0s, but their size and frequency will be completely different. For this reason, it is important that no exceptions are made for any person or device to be provided access to the voice SSID. Failing to do so will only lead to poor voice quality for the VoWi-Fi phones. In the same way, it is a bad idea to deploy smart phones with VoIP apps on a dedicated voice SSID. Voice traffic needs to always have a deterministic advantage over regular data traffic.

Anatomy of VoIP

Voice over IP is a loose collection of proprietary and nonproprietary technologies. Protocols like SIP and H.323 are very commonly used for VoIP. However, they were not explicitly intended for these applications. For example, both can also be used for video conferencing as well. H.323, a common VoIP protocol, was intended for video conferencing. SIP stands for Session Initiation Protocol which adequately describes what it does. And what both of these protocols do is called signaling. Signaling can be described in the most basic terms as the setup and teardown of IP-based streaming communications. Figure 7.5 shows the flow of events in a SIP call illustrated in a packet flow diagram.

This traffic, however, does not contain any voice packets. These are being carried in this instance by another protocol called RTP or

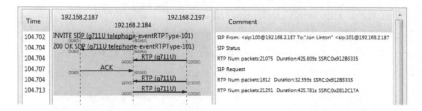

Figure 7.5   SIP_Call_Setup.

Real-time Transport Protocol. Within this protocol the actual codec is visible from a packet analyzer. In fact, it is possible to playback phone calls from a packet capture. This of course is unethical and likely illegal outside of a lab setting. Codecs are the business end of VoIP protocols. They are independent of the signaling and transport protocols such as RTP. Figure 7.6 is an example of a typical RTP VoIP packet.

The highlighted area demonstrates that this RTP stream is using the G.711 codec. The transport protocols need additional protocols like SIP and H.323 to perform the signaling. As a result, there are a tremendous number of combinations of signaling, transport, and codecs available. Not all of these will be common to VoWi-Fi handsets or smart phone apps. Some of the more common are H.323, SIP, SCCP, SVP, which are common signaling protocols, while G.711, G.722, and G.729 are common codecs found on VoWi-Fi phones. Figure 7.7 provides a more complete list of codecs that are used for voice encoding. You will notice that codecs have different transmit intervals (packets per second) and bandwidth requirements (bit rate).

The variation in transmit interval and bandwidth consumption are the result of trade-offs made during the design process. Trade-offs are an ever-present evil of mobile design. Power consumption,

```
⊞ Frame 1716: 214 bytes on wire (1712 bits), 214 bytes captured (1712 bits)
⊞ Ethernet II, Src: d8:d1:cb:74:6a:db (d8:d1:cb:74:6a:db), Dst: Siemens_6e:10:6f (00:01:e3:6e:10:6f)
⊞ Internet Protocol Version 4, Src: 192.168.2.197 (192.168.2.197), Dst: 192.168.2.179 (192.168.2.179)
⊞ User Datagram Protocol, Src Port: ndmp (10000), Dst Port: avt-profile-1 (5004)
⊟ Real-Time Transport Protocol
   ⊞ [Stream setup by SDP (frame 745)]
      10.. .... = Version: RFC 1889 Version (2)
      ..0. .... = Padding: False
      ...0 .... = Extension: False
      .... 0000 = Contributing source identifiers count: 0
      0... .... = Marker: False
      Payload type: ITU-T G.711 PCMU (0)
      Sequence number: 609
      [Extended sequence number: 66145]
      Timestamp: 2863370767
      Synchronization source identifier: 0x70fd4177 (1895645559)
      Payload: fcfafbfb7e7efe7e7d7c7b7a7a7b7e7b7d7b7c7efefe7efc...
```

Figure 7.6   RTP_G711.

	Bit Rate (Kbps)	Sample Size (Bytes)	Sample Rate (kHz)	Sample Interval	MOS	Voice Payload (Bytes)	Voice Payload (ms)	Packets Per Second
G.711	64	80	8	10 ms	4.1	160	20 ms	50
G.729	8	10	8	10 ms	3.92	20	20 ms	50
G.723.1	6.3	24	8	30 ms	3.9	24	30 ms	33.3
G.723.1	5.3	20	8	30 ms	3.8	20	30 ms	33.3
G.726	32	20	8	5 ms	3.85	80	20 ms	50
G.726	24	15	8	5 ms	3.85	60	20 ms	50
G.728	16	10	8	5 ms	3.61	60	30 ms	33.3
G.722	64	80	16	10 ms	4.13	160	20 ms	50

Figure 7.7 Voice codecs.

performance, form factor, screen real estate, and speed are just a few key design parameters that engineers must continually compromise on when designing products. If it was easy, and the laws of physics did not apply, we would all have super computers that are thinner and lighter than today's smart phones. As mobile computing continues to evolve fewer compromises may be necessary to deliver the same services. This is why there are so many codecs. Some codecs require more CPU cycles and thus more power consumption but require less bandwidth. Less bandwidth in turn requires less power consumption. From this list it is important to note transmission interval. The lower the transmission interval is, the more often transmissions occur. Transmissions expend power. Inversely, if the transmission interval is longer, there is a higher likelihood that a retransmission will result in a detectable audible error. There is one other factor that plays into this. Consistent access to the WLAN is very important to VoIP applications. Variations in the delivery time of VoIP packets are called jitter. As the transmit interval decreases, statistically there is more jitter possible.

The Anatomy of Codecs

The human ear is an absolutely incredible part of the body. Small rapid changes in the air pressure are converted into a signal that our brain interprets as sound. In much the same way that the human ear is able to take voices and sounds and translate them to something that can be stored and recalled, coder decoder (codecs) provide a way for computers to emulate their very creators. The digital age was brought about by the creation of the transistor. This single invention created the concept of representing information that was once stored through text from a pen or typewriter in a series of 1s and 0s. Eventually, someone realized the audio spectrum could be represented by 1s and 0s as well. If one tries to represent the complete audio spectrum in a one for one manner it will consume an incredible amount of data. When codecs were first used, data transmission was primarily sent over analog lines that used modulation techniques to try and coerce as much throughput as possible out of the bandwidth available. This functionality came at quite a premium. For example, you could take a pair of cables and transmit an analog voice signal through them or transmit 1200 bits per second. As time passed this was increased to 56 Kbps. This is still below the bandwidth required to transmit even just one VoIP call with the G.711 codec, and even the most miserly codecs like G.729 encode at 8000 bits per second. At the time, it was clearly more efficient to transmit analog voice signals. Once modulation efficiency began to pick up and produce more throughput the idea of transmitting voice over data lines as 1s and 0s became a reality. A great deal of the success of this story is owed to the humble codec. The audible range of the spectrum, the audible range of the human ear, silence, and sampling are key criteria for designing a codec. Let us look at where to start examining and comparing codecs. Harry Nyquist devised a concept that the sample rate needs to be twice the maximum frequency being sampled. This is called Nyquist's rate. The logic behind this is really quite simple. No complex mathematical equations are necessary but an illustration will help to convey this concept. Figure 7.8 shows a generic sine wave with two vertical lines illustrating sample points.

Having a sample rate that is greater than two times the frequency being sampled will only improve the accuracy of the measurement. To get the shape of the waveform you will likely need to sample at greater than 10 times the highest frequency being sampled. This is called oversampling and is a very common practice in audio codecs. The generally accepted range of frequencies accepted as the typical audible range is 20 Hz to 20 KHz. That is an enormous range of values in terms of analog-to-digital conversion. Many codec sample rates are around 8000 sample per second. Based on what we learned about the Nyquist rate we can say that we would need in the neighborhood of $(20,000 \times 2) - (20 \times 2)$ at a minimum to properly sample the entire

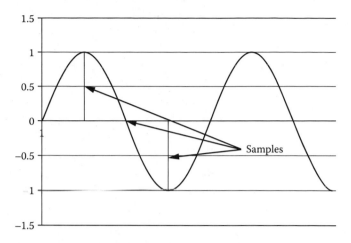

Figure 7.8 Nyquist-2 sample.

audible spectrum heard by the human ear. It would be safe to say we must not need the entire spectrum to be sampled. The 88 keys on a piano produce a range of sound from 27.5 Hz to 4.2 KHz, and the human voice produces a range of frequencies from 20 Hz to 14,000 Hz; however, the majority of the higher energy sound is below 1000 Hz. The generalized voice range is around 500 to 2000 Hz. As a result, it should not be surprising that many codecs use what is called a band pass filter to sample only the 300- to 3,400-Hz range. This sample range has been a longstanding standard. If this range is sampled at 8 KHz it will exceed the requirement of Nyquist's theorem. The codec G.711 uses a modulation technique to quantify each of these samples. There are 8,000 samples per second and each sample is 8 bits, which works out to 64 kilobits per second per stream. The modulation used is call PCM or Pulse Code Modulation. PCM is one of the simplest modulation techniques and is utilized far beyond telephony communications. You could say it is a cornerstone of digital audio. Each individual sample point measures a different frequency within the spectrum range. The intensity of each frequency is translated to a numeric value that is stored as a binary value. And there you have it—the magic of codecs. The number of codecs available is seemingly limitless. Look at the sample rate column in Figure 7.7. We can see that every codec except G.722 uses 8,000 samples. However, G.722 uses twice this value for its sample rate. One would think that oversampling would produce better audio quality; however, there is an

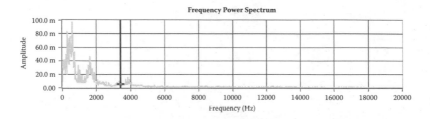

Figure 7.9   Fast Fourier Transform of the human voice.

effect of diminishing returns. As such, the real reason G.722 utilizes twice as many samples is because it is sampling nearly twice as much spectrum. We mentioned that most codecs use a band pass filter to only encode 300 to 3,400 Hz. This is called a narrowband codec. G.722 encodes the range from 300 to 7000 Hz. Surprisingly, this is called a wideband codec. Figure 7.9 shows a spectrum plot of the typical human voice range.

In this plot two things stand out. First, most of the energy is concentrated below 1000 Hz. Second, there is virtually no sound above 5000 Hz, but there is a measurable amount of energy above the 3,400-Hz cutoff of most codecs. Looking back at the codecs chart, you may notice that G.711 and G.722 are nearly identical, with one minor difference in the sample size. The data throughput requirement is the same for both codecs because the modulation used in G.722 is a modified version of PCM called Adaptive Differential Pulse Code Modulation (ADPCM). The important thing to note is that the algorithm is a little more computationally intense for encoding and decoding. When complexity is added to a codec and processing power remains constant, encoding latency will be increased. While a modern laptop computer will have no problem handling the codec quickly, this could add latency to a Wi-Fi handset. Even a VoIP app on a smart phone will require additional latency though possibly less than the Wi-Fi phone, but will increase the power consumption. As such G.711 will introduce latency well below just one millisecond but G.722 could introduce tens of milliseconds of delay. As we will see later in this text 10 milliseconds is an eternity in voice time. Combine this with latency from roaming between APs and this could mean call interruption. As a general rule, when the codec compression or efficiency increases so will the time necessary to compute the signal. This is sometimes described as algorithmic delay.

The sample interval describes how often the codec packages the voice data in a packet. This ranges from 5 to 30 ms. This is not to be confused with the payload, which is the frequency with which the samples are transmitted or packetized. The payload is visible in a packet capture and contains one or more samples. This is usually 20 or 30 ms and can also be expressed as packets per second (PPS). We can walk through an example using the G.711 codec. It has a sample size of 80 bytes, a sample interval of 10 ms, and a payload of 160 bytes. As such, the voice data is continually sampled but packetized in 80 byte chunks every 10 ms. These chunks are bundled together two by two to create a 160-byte payload that is transmitted wirelessly every 20 ms or 50 times per second. There are some newer features being added to codecs through annexes.

It is important to understand that these annexes are not reverse compatible and should be considered as a different codec. This is important for interoperability of phones. For example, G.729B is also referred to as G.729 Annex B. This is not compatible with G.729 or Annex A. Phones negotiate the codec during call setup. Both ends must support the same codec. Typically they will negotiate to the first codec that is supported on both phones. There are three features pertaining to codecs that we need to look at: silence suppression, Voice Activity Detection (VAD), and Comfort Noise Generation (CNG). When two phones engage in a call, each creates a transmit stream destined to the other device. It is intriguing to consider that while both phones are creating voice packets and traffic, only one person is usually speaking at a time. This would seem like a waste of bandwidth. For this reason, many codecs have a feature called silence suppression. By measuring the background noise of a voice call, a noise floor can be established. Silence suppression simply adds an algorithm to determine whether a sample is sufficiently above the noise floor in order to transmit it. If the algorithm determines that it is silence then it will not transmit it. You may have heard this in a call without knowing exactly why. Perhaps after a long chat with an old friend you both reach the inevitable pause. However, one of you interjects as it would seem that the phone call has cut out. It turns out that the call has not been cut short but instead just gone incredibly quiet. The silence suppression solves the problem of wasting bandwidth but creates confusion and an unnatural experience.

Our brains may associate complete silence with a call disconnect or other interruption. This is why someone came up with CNG. CNG will try to algebraically estimate the background noise after taking measurements. The only transmissions that are created are updates to the background noise "description" instead of the actual noise. This is synthesized on the other end and *voila* we have a normal sounding call and bandwidth savings. There is a small penalty in creating the background noise as this will take a few processor cycles to create. However, there is a big benefit in eliminating a transmission from the radio. This will positively affect the battery life of the phone and reduce the bandwidth consumption. The final feature is VAD. This is similar to silence suppression except in just the opposite manner. Instead of detecting samples of silence and omitting them, VAD detects samples of nonsilence and transmits them. One problem with both VAD and silence suppression is making sure that the correct portions of an actual call are heard. Again, the phone's processor will need to consume additional power to correctly validate voice activity, but has the added benefit of reducing bandwidth consumption as well as battery consumption from radio transmissions.

Wi-Fi can add some unique challenges to VoIP and for this reason it is recommended to choose a codec like G.711 that is widely used and freely available for mobile applications. This will help ensure interoperability and minimize additional engineering considerations. Whether this is leveraged by a smart phone or a dedicated Wi-Fi handset, G.711 has proven to be the gold standard for any initial VoIP system deployment.

Proprietary Protocols

Regardless of the codec used, it is likely that the transport protocol will be the same even for proprietary protocols. RTP is very much the de facto transport protocol for voice. SCCP or "Skinny" and SVP are proprietary protocols created by Cisco and Spectralink, respectively. SCCP only implements call signaling much like SIP and H.323. The transport and codecs are standard fare. SVP is completely proprietary. The signaling, transport, and codecs are all included in the protocol. At the time it was created, SVP solved a big problem that existed for wireless LANs. The original 802.11 standard did not have any QoS mechanisms. SVP provided a way to provide voice communications over Wi-Fi without the requirement for traditional QoS mechanisms. SVP relies on gateways to provide timing or scheduling of packets to work optimally over Wi-Fi. The details of the how this protocol works have not been officially published, but by looking at the protocol and packet analysis we may be able to confirm a few things. First, SVP packets are sent in 30 ms intervals. And second, the extra code in the SVP capable APs allows the SVP packets out into the air without the standard random back off timers. This extra code is called Spectralink Radio Protocol (SRP). This is a bit of "cheating" by the protocol but that isn't necessarily bad as long as one is aware of the variance. By cheating the system, SVP limits the effectiveness of Wi-Fi standards like WMM and 802.11e. These are covered in detail later in this chapter. Now that there is a standards-based approach to QoS for Wi-Fi, it is not necessary for SVP to be used, though many organizations still use it. One of the added benefits of protocols like SVP is that it is not necessary to create wired and wireless QoS policies.

Skype is another proprietary protocol. Skype has created several protocols and codecs for home use. Mass adoption of consumer devices in the healthcare space requires engineers and stakeholders to understand this technology's merits and shortcomings. Skype began with the idea of providing free long distance calls over any Internet connection. Its developers have since invented two additional codecs and taken aim at mobile devices. Like SVP Skype attempts the impossible, of providing toll quality voice calls over IP communications without QoS. In traditional networks this is not as big a deal as one might think. While implementing a QoS policy on the switches and routers in a network is generally a good practice, it is not a necessity. Designing VoIP networks on routers and switches has a lot to do with moving bottlenecks. When bottlenecks cannot be eliminated or moved, a solid QoS policy must be put into action. Unfortunately, Wi-Fi APs completely change this paradigm due to the medium access method. To be able to provide reliable scalable VoIP, even technologies like Skype, it is necessary to use WMM or 802.11e Wi-Fi QoS. This will often necessitate a wired QoS policy as well. With that in mind it is unlikely that technologies like Skype will be able to dominate the mobile enterprise VoIP market. That said, Skype has become very popular among patients and patient guests. A hospital's guest network might as well be an ISP as well as a PBX. It is important to identify early on in the design phase of the WLAN whether or not applications like Skype for patients will be desired. Providing a great experience for guests and patients with proprietary apps is a difficult task. Even a well-designed network may not be able to provide the best experience with protocols like Skype.

Wireless Arbitration

Wi-Fi Multi Media (WMM) is a standard developed by the Wi-Fi Alliance. WMM adopts most of the IEEE 802.11e standard that is now simply part of the 802.11-2012 standards. WMM modifies the way in which APs and clients access the wireless medium. Since Wi-Fi is a shared medium, clients and APs must contend for the right to transmit. This process is called arbitration. The 802.11 designers came up with a clever way to address the problem of multiple devices gaining access to the same medium. A BSS or basic service set is used to describe the relationship between a client and an AP. The BSS has rules that clients follow on the honor system. As mentioned before regarding the SVP protocol, sometimes the rules are bent to favor one over the other. By using a set of rules and random timers every device is granted roughly equal access to the network. While there can be brief (milliseconds in Wi-Fi time) periods of inequality of access, on average over a longer duration the access should be the about same if every client and AP is using the same rule set. This is fine for data, but there is a problem with it. For starters, every wireless frame must contend for access. There is no guarantee that the voice packets will maintain their 20 or 30 millisecond transmission interval reliably. In voice communications this variation in the delivery time of packets is called jitter. This is perceived by the human ear as a stuttering or cutting out of a voice sometimes in the middle of a word. These experiences are also described as artifacts. Whatever you may call them, they will certainly frustrate the individuals that experience them. This is why 802.11e was developed. The IEEE 802.11 task group set out to design a QoS method into the 802.11 standard. This standard is now simply a part of the 802.11-2012 standards and is no longer referred to as just 802.11e. In addition, the Wi-Fi Alliance has adopted the majority of the 802.11e standard to create what they call WMM. Actually, the Wi-Fi Alliance created the WMM certification prior to 802.11e being ratified. Wi-Fi networks operate under a set of rules called Distributed Coordination Function or DCF. This construct provides fair access to the wireless medium. As we have already mentioned sometimes we do not want fair access. This is where EDCA comes in. EDCA stands for Enhanced DCF Channel Access and extends the ruled defined in DCF. Before we can discuss EDCA we will need to talk about DCF. DCF is the fundamental basis for Wi-Fi as we know it. To properly understand DCF there are a number of concepts and definitions we need to look at.

- Interframe spacing
- Time slots
- Contention windows
- Random back off

Interframe spacing is designed to prevent wireless communication failures. In other words this helps clear the air before transmitting. First, is the SIFS or short interframe space. A SIFS is 10 μs for 802.11b/g/n or 16 μs for 802.11a/n. SIFS are only used prior to Wi-Fi frame transmission after the medium has been reserved. The purpose of the IFS is to prevent interference between transmissions but not waste any of that precious airtime. A SIFS is the second shortest IFS next to the RIFS. A RIFS has the same aim as the SIFS with the only difference being the duration. A RIFS is only 2 μs in duration. One more note about RIFS; they are completely imaginary which makes the SIFS the shortest real interframe space, and it will likely remain that way even for new standards. The wireless medium is reserved by the devices waiting for a DIFS or an AIFS. A DIFS is equal in time to one SIFS plus two slot times, which will be different for 802.11a/n and 802.11 b/g/n. A slot time is 9 μs for 802.11a/n and 20 μs for 802.11b/g/n. Therefore a DIFS for 802.11a/n and 802.11b/g/n is 34 μs and 50 μs, respectively. DIFS are used to gain access to the wireless medium in a BSS but AIFS are used to gain access to the wireless medium in a QBSS.

We discussed the concept of a BSS earlier in the chapter. By 2002, several Wi-Fi handset makers had emerged on the market. In fact, the telecommunications juggernaut Cisco purchased a promising VoIP company named Selsius in 1998 and later began to add Wi-Fi hand-sets to its product line a few years later. However, there was a problem with VoIP running on Wi-Fi at that time. Prior to September 2004, Wi-Fi did not have any way to prioritize the voice traffic over the data traffic. The end result of course was 802.11e from the IEEE, WMM from the Wi-Fi Alliance, and ultimately the QBSS. Many individuals I speak with who are new to wireless don't understand that it only allows one client to transmit at a time. This is called half-duplex communication and uses something called CSMA-CA. Carrier Sense Multiple Access with Collision Avoidance sounds so much better as an acronym, which is why this is the last time I will fully write it out. When you make a call on your cell phone, your voice is modulated over one set of frequencies while the voice you hear is modulated over a completely separate set of frequencies. Wi-Fi uses the same set of frequencies to transmit and receive. While this adds to the spectral efficiency and duty, it does make real-time bidirectional communication a bit of a challenge. Anyone who has worked with Ethernet communications is more familiar with CSMA-CD, which is CSMA-Collision Detection. CSMA-CD is typically used in wired network communications and can detect collisions because it will notice a difference in the voltage. Wires are not known to grow and shrink thereby changing their resistance and end-of-line voltage during the course of the normal day. The signal received at a client device changes by the microsecond. For this reason it is important to avoid collisions in wireless communications. What is important about the difference in these two technologies, which despite having one letter difference could not be more far apart, is that CSMA-CD allows two devices to talk simultaneously and CSMA-CA only allows one. Perhaps if you are especially proficient in wireless technology you will be screaming at these pages that 802.11ac will allow multiple transmissions at once. You are correct but, as of this writing, 802.11ac has not had widespread adoption, though we suspect it will. Now, we were talking about the QBSS. One of the biggest differences between a BSS and a QBSS is the way in which devices access the wireless medium. Wireless clients use something called interframe spaces to help prevent interference caused by multipath transmissions. Multipath can be thought of as echoes. This is not completely accurate but is a good illustration. By waiting in interframe space, the air is cleared out of any reflected RF. There are five different interframe spaces and each serves a different purpose.

- SIFS: short interframe spacing
- DIFS: DCF interframe spacing
- RIFS: reduced interframe spacing
- AIFS: arbitration interframe spacing
- EIFS: extended interframe spacing

The duration of each of these interframe spaces varies based on several factors such as the 802.11 standard in use. They are basically broken down into three categories. The first contains the DIFS and AIFS, which are used to gain access to the wireless medium. The second is the SIFS and the RIFS, which are used between frames after a SIF or AIFS has been used. Lastly, an EIFS is used when a collision is detected. An EIFS has a substantially longer duration to ensure that the transmission is received without a collision. To be approximate, an EIFS is equal to a SIFS plus a DIFS plus the preceding ACK. This will vary based on the PHY and minimum basic rate supported. As we mentioned prior, a DIFS has the same duration as a SIFS plus two slot times. By contrast, a QBSS uses a matrix of values to determine how long to defer transmission based on which Wi-Fi standard is used and which priority queue is used. The AIFS duration is the AIFSN times a slot time plus a SIFS. The AIFS number is differentiated between the four different QBSS priorities: background, best effort, voice, and video. Figure 7.10 shows the different AIFS numbers mapped to their priority queues and the AIFS duration for each.

Each priority has a range of possibilities that are different from the next queue with the exception of the best effort and background queues. With this design it is possible to get a statistical advantage from the voice queue over the video and the video over the best effort

Priority	AC	802.11b/g/n			802.11a/n		
		CWmin	CWmax	Time	CWmin	CWmax	Time
Highest	Voice	3	7	70–150	3	7	43–79
	Video	7	15	150–310	7	15	79–151
	Best Effort	15	1023	310–20470	15	1023	151–9223
Lowest	Background	15	1023	310–20470	15	1023	151–9223

Figure 7.10   AIFS times.

and background queues. As you can see the priority order from lowest to highest is background, best effort, video, and voice. What happens when two voice packets arrive at the the access point at the same time? There is another mechanism that assists in avoiding collisions, called the random back-off timer that occurs during the contention window. This is the computer equivalent of "rock paper scissors." Based on what priority the traffic is defined as, the AP will randomly select a different number to count down from for each frame. This whole process allows for higher priority traffic to have a higher statistical probability to gain access to the medium. There are no guarantees that one traffic type will get priority over another. It is possible for lower priority traffic to occasionally transmit before higher priority traffic. On average over a longer duration of time, the traffic should follow the traffic priorities specified.

Troubleshooting VoWi-Fi

It is very likely that during the course of the deployment and fine tuning of a VoWi-Fi solution it will be necessary to troubleshoot a problem. The best place to start troubleshooting a VoWi-Fi problem is to validate the RF quantitatively and qualitatively. In order to troubleshoot a WLAN, there are some tools that will be a priority to purchase such as Wi-Fi analysis software, spectrum analyzer, site survey software, and possibly VoWi-Fi analysis software. The cost of these tools ranges from $100 to $10,000. On the other hand, wireless consultants can cost hundreds of dollars per hour. Many organizations may opt to train their own employees to use these tools to troubleshoot. One method of doing this is to walk around the affected area with a Wi-Fi analyzer or preferably site survey software. A Wi-Fi analyzer typically uses a standard wireless card with a special driver that allows the card to be placed into monitor mode. Site survey software will work in a similar manner but may record less information. In this mode the WLAN card will scan through all the channels and record a lot of information. One very common measurement is the RSSI, often measured in dBm or decibel milliwatt. The first measurement to validate is the minimum RSSI or SNR measured in the facility. It may be fine if the signal drops to –70 dBm and it may be necessary for the signal to be at –65 dBm to provide proper coverage but –67 dBm is a good starting point. What will really determine if a voice packet is properly received is the SNR value. Performing a site survey with a spectrum analyzer or simply taking spectrum analysis measurements allow the SNR to be recorded accurately. The site survey software has the benefit that it can provide a location pinpoint with the SNR measurement but it also can cost a lot more. Ultimately, RF problems will likely correspond to a locality or set of APs. Generally speaking, a low SNR will be attributed to one or more factors, such as:

- AP transmit power set too low
- AP transmit power set too high
- Non-Wi-Fi interference

When the AP transmit power is set too low, this can create "dead spots." It is a temptation to set all the APs to the maximum configurable output power to resolve this problem. Well, not so fast. If the signal is too hot, we will likely see co-channel interference. Co-channel interference usually comes from APs that are on the same channel that are either too close, have the transmit power set too high, or both. This effectively raises the noise floor, which lowers the SNR value. When this happens retransmissions are likely to occur which will degrade voice call performance. During the design phases it is important to configure adjacent APs on different channels. For the 2.4-GHz spectrum this is difficult since there are only three non-overlapping channels to use. In the United States, these are typically 1, 6, and 11. The 5.0-GHz spectrum has up to 24 non-overlapping channels to plan with. As we mentioned earlier in the chapter, roaming is affected by the number of channels used. Just because there are 24 channels does not mean that you should use them all. Channels 1, 6, and 11 have a little bit of spacing in them when using OFDM; however, the 5-GHz channels do not have spacing between adjacent channels. Wi-Fi radio amplifiers create side lobes in transmissions that can interfere with adjacent channels. For this reason it is recommended that adjacent APs in the 5-GHz channels be configured with one cell of spacing for APs operating on adjacent channels. It is recommended that APs using the same channel have two cells of spacing between them. It is important to use the actual VoWi-Fi phones being used as a benchmark for RSSI measurements. Every radio hears RF signals differently. Some radios will not hear the signal as loudly as the Wi-Fi analyzer will and may even amplify the signal. Low noise amplifiers, which attempt to increase the intended signal without increasing the noise floor, have greatly improved since the earlier days of Wi-Fi; however, they will certainly still introduce additional noise into the original signal. What this means is that not only will the signal be amplified but so will the noise level. Try to imagine that you have a record that is in bad condition. And let's say that the music is a quiet concerto from Bach or Beethoven. Slightly annoyed that you can barely hear the music while trying to relax you attempt to turn up the volume. To your surprise this results in the hissing and popping of the wrecked record being amplified as well. Amplification works by adding noise to a signal to make it greater in amplitude. Unfortunately this also raises the noise floor. Every Wi-Fi chip manufacturer publishes specification sheets on their chips which describe what data rate they should be able to receive and transmit at various SNRs. Most Wi-Fi analyzers will display a noise level value in their measurements. The problem with this value is that it is not a true measurement. The noise level can only be measured by a spectrum analyzer. The noise value is calculated and so is the SNR. Since Wi-Fi radios only understand Wi-Fi and cannot measure things like the noise floor, this begs another question. What about interference sources like Bluetooth, microwave ovens, and weather radar blasts? As you can guess, it does not understand these either. Why would we ever bother to validate that a Wi-Fi network supplies an RSSI of −67 dBm then? This has to do with RF noise. RF noise is everywhere and is mainly caused by heat energy mostly from the sun for us earth dwellers. The short answer then is that if the only thing impeding your Wi-Fi performance is the noise floor, then −67 dBm will be sufficient to provide error-free transmissions. This measurement is best taken from a VoIP phone in site survey mode which will provide the RSSI value from the phone's perspective. Keep in mind that this is a starting point and may not always be sufficient. Some phones work perfectly fine at −72 dBm while others work poorly at −65 dBm. What if there is interference? If that is the case, the Wi-Fi analyzer will not be able to identify this from its noise floor measurement like a spectrum analyzer would. But there may be some tell-tale signs of a problem. One such measurement will be the retransmissions and errors. It is important to pay attention to where this measurement comes from. There are at least three places this can come from: the client, the AP, or a measurement tool such as a sensor or Wi-Fi analyzer. When a client or AP does not receive an ACK frame from its last transmission it will retransmit the frame until it receives an ACK or exceeds its timeouts. The problem is that if the device receiving that transmission cannot hear the frame at all, it will not increment the error or retransmission values. Retransmission is a part of Wi-Fi but it needs to be minimized in order to achieve a high performance level. Generally, measuring retransmissions in between the client and AP may provide false information. Ideally, the transmission would be measured at both the client device and AP. This more closely portrays the experience from each end's perspective. On occasion it may make sense to capture the traffic somewhere in between the two end points as this will normalize the interference level. There are a great number of tools available commercially to assist in troubleshooting VoWi-Fi. One problem with these tools is that they will likely only validate whether the WLAN performance meets the requirements for VoWi-Fi. When using a VoWi-Fi analyzer, it is important to keep in mind the prior discussion about signal levels and retransmissions. The tools are only automating test and analysis, but they need good data to work with. On way to increase the likelihood of capturing good data is to use multiple wireless adapters in a USB hub. This allows the person doing the analysis to capture all the data from several APs at once. This type of approach is a necessity for performing roaming analysis. Wi-Fi roaming should occur in under 50 ms. Wi-Fi analysis tools typically listen on a channel for 100 to 250 ms before moving on to the next channel. It is very likely that important information will be missed with the channel hopping. At least two Wi-Fi adapters are needed to successfully capture a roam. Each adapter is set to listen to a different channel. This ensures that none of the information is missed.

Many Wi-Fi analyzers have the ability to provide an R-factor and Mean Opinion Score (MOS) to describe approximately how well the VoIP is working on the wireless LAN. There are some nuances to these measurements. Let's start by looking at the MOS. The MOS was developed to quantitatively measure what humans perceive as acceptable communications. The MOS is measured on a scale of 0 to 5, with 5 being excellent and 1 being poor, and it works exactly as it is described. It is an average or mean of human subjective opinions.

User Opinion	R-Factor	MOS
Typical G.711	93	4.1
Best	90–100	4
Good	80–89.9	3
Some users dissatisfied	70–79.9	2
Many users dissatisfied	60–69.9	1
Nearly all users dissatisfied	50–59.9	0

Figure 7.11   R-factor, MOS, and user opinions.

Each codec has a maximum achievable MOS. For example, a common codec like G.711 is expected to have a MOS no higher than 4.1. This is considered good audio and sufficient for voice communications. In fact, any score equal to or greater than 4.0 is considered telephony grade audio. Analyzers will approximate the MOS values based on several factors such as retransmission and error rates. These figures are a good way to numerically standardize testing procedures but have not replaced the judgment of actual people. In other words, it is important to validate a wireless voice network qualitatively as well. Delivering packets is part of the equation, but the other part is in what is in the packets. For example, an analyzer may score a wideband codec identical to a narrowband codec, but a human ear will be able to tell you that the wideband codec sounds better than the narrowband. The MOS is important to keep in mind when testing, validating, and troubleshooting, but ultimately provides qualitative estimates from quantitative measurements.

The R-factor is a calculated measurement of the quality of a telephone call. The range of values is 0 to 100 but 50 to 94 is the generally used range. Figure 7.11 shows the correlations between R-factor, MOS, and user opinion.

The R-factor is based on what is called the E-model. The E-model was developed by the ETSI to quantify telecommunication impairments. The R-factor is defined by the equation

$$R = R_0 - I_s - I_d - I_e + A + W$$

where:

R_0 = noise floor (audible not RF)
I_s = simultaneous impairments
I_d = delay impairments

I_e = equipment impairments

A = advantage factor

W = can be added to account for wideband codecs

The MOS and R-value are useful for measuring the quality of the network environment, but other factors can weigh in and create an undesirable experience even though the measurements indicate otherwise. One of the biggest problems with VoWi-Fi is the difficulty with roaming.

Roaming

All devices must perform two operations to join a Wi-Fi network: authenticate and associate. When a device roams it performs similar functions. There are four Wi-Fi frames involved in joining a network: authentication request, authentication response, association request, and association response. In order to roam to another AP a phone will need to perform scanning, probing, and then reassociation. During the scanning phase a phone will use time slices in which it is not transmitting or receiving wireless frames. It is looking for beacons from other APs, which are sent out at intervals of 100 time units or approximately 102.4 ms. Hence, the phone may need to listen for up to 307.2 ms to hear APs on channels 1, 6, and 11 in the 2.4-GHz frequencies.

In the 5-GHz spectrum this is even more difficult. If we assume that the UNII-2 bands are omitted then we only have eight channels to scan. This means that the phone will need up to 819.2 ms to scan the channels. If all the 5-GHz channels are enabled this could take nearly 2.5 seconds. This can be a timely operation and therefore must occur continually to be effective. Dedicated Wi-Fi phones will allow an administrator to select which channels to scan on. It is usually a best practice from all vendors to use the 5-GHz channels exclusively. This is not a bad thing as the 2.4-GHz spectrum is increasingly crowded. In fact, it shares spectrum with at least two other protocols, namely Bluetooth and Zigbee.

Spectrum plays an important role in the roaming process. Scanning allows the client to get a rough estimate of the available APs in the area. A Wi-Fi phone will send out a probe request frame. Any AP in the broadcast range of this phone that would like to respond can do so with a probe response. The probe response tells the phone a number of things about the network, such as the SSID, supported data rates, and even the BSS load. The phone uses these elements and fields to decide where to attempt a roam. When the phone selects an AP it will send the new AP an authentication frame. Once authenticated the phone can now send a reassociation request to the new AP with the MAC address of the current association. Most enterprise WLAN vendors have code written in an AP or controller that allows APs to share client connection information to aid in the roaming. For open unencrypted networks, like a hotspot or guest access, this is where the story ends. In a healthcare setting, it is more likely that the voice network will be encrypted with a security suite like WPA2-PSK. This is covered extensively in the chapter on wireless security. Encryption keys are designed to obscure the contents of a data frame and they are continually updated. When a wireless network is encrypted the clients and APs must share and negotiate their keying information during a roam. This occurs during the four-way key exchange. This key exchange creates additional overhead and delay and must be factored into a LAN design. Preshared key encryption suites do not add significant enough overhead to prevent successful roams. However, adding authentication such as 802.1x will create more roaming delay and likely cause voice degradation during roams. 802.1x is more commonly used on laptop computers than on Wi-Fi phones.

There is a growing trend to attempt to use smart phones as VoIP handsets. There are significant scalability issues with handing out preshared keys to everyone with a smart phone. For that reason it is likely that 802.1x authentication will need to be used for these types of multiservice devices. With these limitations the outlook does not look good for VoIP on smart phones. There are some glimmers of hope with 802.11k and 802.11r. These two standards are designed to help expedite the roaming process for devices like smart phones. 802.11k is referred to as radio management and 802.11r is sometimes also called Fast BSS transition. 11k aims to improve the scanning and probing processes that occur continually prior to roaming. This is achieved by including which channels are in use by neighboring APs. A phone or any client for that matter can use this information to only scan and probe on the suggested channels. Let's say for example that a phone is notified that there are neighboring APs on channels 36, 44, 149, and 157. As we discovered earlier, it could take as long as 2.5 seconds for a phone to hear beacons from every channel in the 5-GHz spectrum. If it knows it only needs to listen for 102.4 ms on four channels this is a 600 percent improvement in scan time. Unfortunately probing efficiency will not improve as the scanning will define where to probe typically. 802.11r will greatly improve the roam times of 802.1x authenticated wireless. In fact, roam times should be in line with those of preshared keys. The manner in which this is accomplished is actually quite logical and simple. The problem with WPA1/2 with 802.1x authentication roaming is the quantity of wireless frames that must be exchanged. 802.11r reduces the number of exchanges by piggybacking the four-way handshake with the authentication and reassociation exchanges. The Wi-Fi Alliance is calling this WMM Voice-Enterprise. 802.11r will play a critical role in enterprises adopting VoWi-Fi on smart phones, tablets, and laptops. To date very few phone and more importantly WLAN vendors have implemented 802.11r achieved Voice-Enterprise certification. New technologies are slow to be adopted by enterprises.

Two other solutions for faster roam times are Opportunistic Key Caching and Pre-Authentication. Both of these were created with the 802.11e standard. Opportunistic Key Caching allows a client to roam without the need to derive new encryption keys. It does this by sharing the keying information among all the APs and controllers within a mobility domain. This reduces the key exchanges from reassociation, EAPOL, and the four-way handshake to just the reassociation and four-way handshake. Opportunistic Key Caching can reduce the number of frames necessary to roam by as much as 50 percent. Pre-Authentication does not reduce the number of frames exchanged, but rather reduces the number of frame exchanges during the actual roam. A client can perform the necessary 802.1x authentication step with other APs in the vicinity. In doing so, the client device will only need to perform the reassociation and four-way handshake exchanges. There is a caveat to both of these solutions. This functionality must be supported on both the client device and the WLAN. These are not supported on very many client devices. As of this writing Cisco is the only vendor supporting 802.11r and the iPhone with iOS 6 is the only phone supporting it.

Voice over Wi-Fi has found its way into many organizations. While it may seem simple to flip the VoWi-Fi switch on your wireless LAN gear, hopefully you now have a deep respect for the engineering effort that goes into designing and managing a voice network. It is of the authors' opinion that there are better ways to deliver mission critical voice communication to hospital staff. While it is technically feasible to provide this technology, it certainly does not come without cost. As the current and emerging standards become more widely adopted, VoWi-Fi may become a more trivial task. Presently, this is far from the case. I do not believe VoWi-Fi will fizzle out; in fact, it would seem that it will become integrated into applications and devices. And when it does, hopefully you will be prepared to handle the task at hand.

8

REAL-TIME LOCATION SERVICES

Real-Time Location Services (RTLS) offer an array of enterprise solutions to track and locate assets, including devices and people within the hospital facility. RTLS has gained tremendous traction in the healthcare industry. With the number of use cases it is not surprising hospital staff are using these systems to create efficiencies to save time and money. This chapter provides an overview of RTL that will be useful whether you are an engineer who may be implementing RTLS or an executive who would like to understand more about how to use RTLS to create efficiencies.

The clinician of today has an insurmountable amount of work to manage. Let's look at the job responsibilities of a hospital nurse, who often provides the most direct patient care. He or she is responsible for carrying out doctors' orders, taking vitals, administering medications, acting as liaison between doctors and patient families, and supervising the use of medical equipment. A properly configured RTLS system can minimize the task of tracking down medical equipment. Device hoarding is a term known in hospitals that refers to the hiding of devices for use because they can be hard to come by when needed most. Hoarding is especially problematic to the biomed folks who have to regularly maintain and upgrade these devices. RTLS can not only show the nurse where the medical device is located, but it can show the status of the device. Is the machine ready for use or in need of service? The ability to quickly identify and track equipment needed to administer patient care is tremendously valuable. We will discuss a number of other use cases providing similar advantages later in this chapter.

For the purposes of this book we will focus on RTLS as a local positioning system used inside hospital facilities. Tracking is achieved by attaching a transmitter, also referred to as an RTLS tag, to the desired asset. The RF receiver feeds transmitter information to software that provides a graphical interface to monitor status and location. There are a number of RTLS technologies that use different wireless mediums to communicate. Some of these systems will directly interface with the Wi-Fi network and others use alternate wireless technologies. This section will highlight a few different mediums to provide RTLS. Many of these technologies can be installed in combination with Wi-Fi to increase location accuracy.

RTLS Technologies

ZigBee

ZigBee is a wireless technology developed as an open global standard to address the unique needs of low-cost, low-power wireless networks. Zigbee is built upon the IEEE 802.15.4 standard operating in the 900-MHz, 868-MHz, and the top range of the 2.4-GHz bands. ZigBee operates using low-power transmission and uses mesh technology to communicate large distances through intermediary devices. In the 2.4-GHz band this technology peaks out at 250-Kbit/s data rates and is ideally suited for intermittent data transmission for location tracking. ZigBee tag batteries last for an extended period of time creating cost efficiencies. The receivers for this technology are often plugged right into a wall outlet. As a result ZigBee is scaled fairly easily (http://www.zigbee.org).

Wi-Fi

Most hospital facilities have at least some Wi-Fi infrastructure that may be utilized, or expanded, to facilitate RTLS transport. Leveraging hardware already installed may provide cost savings. RTLS vendors have two different ways to utilize Wi-Fi technologies. Many vendors began by utilizing an overlay sensory network acting as the receiver. Just like the WIPS system, discussed in Chapter 10, the main advantage of this overlay type of infrastructure is the offload of processing load from the access point serving client devices. As RTLS has evolved, many systems now use the access points themselves as the receivers. Leveraging Wi-Fi makes sense but keep in mind that accuracy depends on how close the receiver is to the tag. As a result room-level accuracy may be difficult to achieve without using another system to improve accuracy. The integration of Wi-Fi with other RF technology allows for dramatic improvements in accuracy where needed.

Infrared

Infrared (IR) systems use beams of infrared signals similar to common television remote controls. IR can significantly improve the location accuracy to less than 1 meter. Infrared light is fairly susceptible to obstruction, but using narrow beams of IR can help minimize these challenges. The concept of an exciter occurs when an infrared tag crosses a choke point or is in close enough proximity to trigger or "excite" the tag. This triggers the tag to send a communication to the Wi-Fi system. This process provides for a much more accurate location. These technologies are often battery powered making them easy to install; however, be sure to budget the cost of battery replacement accordingly.

Ultrasound

Ultrasound devices are used to detect objects and measure distances by using ultrasonic waves. In RTLS, an ultrasound signal is transmitted that communicates its location to microphone sensors. The wavelengths are very short and thus signals are confined to a room or other facility boundaries like walls. Advancements in directional microphone technology continue to improve ultrasound accuracy down to inches.

How RTLS Works

The measurement of received signal strength, time of arrival, difference in time of arrival, and angle of arrival are methods of calculating position. Most technologies use a combination of these methods to determine location. The next section in the chapter will touch on the basic physics of these concepts.

Received signal strength (RSS) is a way to calculate the distance from transmitter to receiver by using signal strength. Using the RSS from the transmitter, the distance between it and the receiver can be estimated using power readings from multiple receivers. There are

$$\text{Intensity} \propto \frac{1}{\text{distance}^2}$$

Figure 8.1   Inverse square law.

many factors that affect the RSS, making this type of measurement highly susceptible to error. Many obstructions, such as walls or lead lining, can distort the estimated range readings leading to varied location estimates. The inverse square law, which can be used to calculate the distance between the transmitter and receiver, is illustrated in Figure 8.1.

In wireless terms the RSS intensity is inversely proportional to the square of the distance from the receiver. The distance a device is from the receiver can be calculated predictably when it is close to the receiver. As the device moves farther away the accuracy diminishes quickly.

A simple log-linear path loss model to calculate the distance between transmitter and receiver is illustrated in Figure 8.2.

Another way to calculate location using RSS is called RF fingerprinting. RF fingerprinting is the process of mapping x/y coordinates on each floor of a hospital by recording consistent measurements across the facility. Changes in the environment may change the baseline fingerprint data, requiring measurements to be done again and again. This process can be time consuming and expensive. Utilizing other methods of location calculation in combination with RSS may be the best approach.

Time of arrival (ToA) is the travel time from the transmitter to the receiver, sometimes referred as time-of-flight (ToF), and can be used to measure the distance between the two. In order to properly locate a device with ToA, there must be at least three sensors. When the distances from three different sensors are known, the location can be found at the intersection of the three circles created around each sensor, with the radius being the distance calculated (for more information, see http://www.wpi.edu/Pubs/E-project/

$$RSSI = 10\alpha\log(d)$$
$$d = 10^{(RSSI - RSSI_{calibration})(-10\alpha)} + d_{calibration}$$

Figure 8.2   Distance calculation.

Available/E-project-042811-163711/unrestricted/NRL_MQP_ Final_Report.pdf). ToA requires synchronization between all nodes in the network to function. This may be difficult to achieve depending on the technology used.

Time difference of arrival (TDoA) uses multilateration, or hyperbolic positioning, to locate the tag. It is very similar to ToA in that it uses the travel time from the transmitter to the receiver to measure distances. The difference in travel times from each sensor is used to find the distance between each sensor. This results in several hyperbolas, the intersection of which is the location of the transmitter (more details are provided at the URL cited above).

Angle of arrival (AoA) is defined as the angle between the direction of an incident wave and some reference direction, which is known as orientation. Orientation, defined as a fixed direction against which the AoA is measured, is represented in degrees in a clockwise direction from the north (see http://www4.ncsu.edu/~mlsichit/Research/ Publications/aoaLocalizationSecon06.pdf). A disadvantage of using AoA is the need for an antenna array for each node and it is highly susceptible to multipath environments.

Architecture

The proper design to facilitate RTLS has been highly debated in the last few years. Defining the best architecture is difficult. The best design is going to follow the general recommendations provided below and incorporating the vendor best practice for the chosen RTLS vendor. When using Wi-Fi the location of the tag is in one form or another calculated by its distance from the closest receiver. The closer the tag is to the receiver the more accurate the location. This means the more receiver devices the better the accuracy. Placing access points all around the perimeter of the facility is an option. It is especially important to place access points outside the hallway. Without doing so location calculations will likely represent devices in the hallway rather than the actual room. We have seen designs that doubled the access point density. Although this will improve accuracy it is important to minimize adjacent and co-channel interference by carefully planning power and channel plans.

ISO/IEC Standards

RTLS technology is standardized by the International Organization for Standardization and the International Electrotechnical Commission, under the ISO/IEC 24730 series. In this series of standards, the basic standards ISO/IEC 24730-1 and 24730-2 cover RTLS. More information can be found at http://www.iec.ch/.

Different Types of Transmitters

There are two primary types of RFID transmitters, active and passive. Both tags have advantages and disadvantages.

Active RF tags are battery powered and send active signals. Active tags also contain onboard electronics enabling the device to transmit to the reader on its own. These tags can operate up to 300 feet or more. Depending on the environment they will last from three to eight years. Active tags can also support more advanced security configurations that may be required for some applications.

Passive RF tags contain a chip that absorbs radio energy from the reader. These tags do not contain a battery and rely on RF energy to enable signaling. This signaling is achieved by passing RF energy through the tag antenna creating a magnetic field. This magnetic field is what powers the tag. As a result the range on passive tags is significantly less and varies around three meters. Passive tags last for a number of years before needing to be replaced. Some last longer than 10 years.

Both types of tags can store data but active tags provide superior data storage size and other features. However, active tags are significantly larger in size and more expensive (Figure 8.3).

Applications

There are a number of very exciting hospital applications for RTLS. This section will highlight the application and use cases.

Asset Management   The concept of context awareness is a key factor to expanding asset management beyond simple location. We discussed nurses needing to find medical devices for patient care. A context

Figure 8.3 Active tag. (Photo courtesy of Stanley Healthcare.)

awareness system can not only identify where the device is but also provide critical status information. An intravenous pump continuously cycles between different statuses, including in use, soiled, cleaning, available, and needing maintenance. A properly configured system may show the nurse only the location of where an available usable unit is on the floor. The biomed staff may want to see a listing of all the devices needing disinfection. This kind of enhanced visibility will dramatically improve workflows in the modern hospital. Context awareness is now integrating location with utilization metrics. This means the pharmacy not only knows where the pumps are located but also how much of a specific medicine they have deployed. Figure 8.4 shows the location of a soiled pump.

Equipment Rentals Hospitals routinely rent equipment to support patient care needs which can add up to be very cost prohibitive. By tracking the movement of all devices and analyzing historical data it

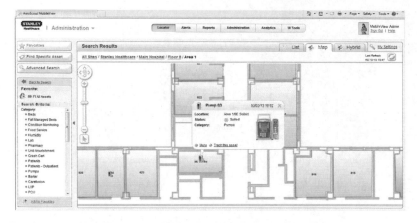

Figure 8.4   Location of a soiled pump. (Photo courtesy of Stanley Healthcare.)

can be determined that a particular area of the hospital is underutilizing a device where another unit may have a shortage. This type of tracking provides insight to reduce overall rental expense.

Shrinkage   Lost or stolen equipment is a constant challenge in the hospital environment. Tagging devices that regularly go missing can generate data used to minimize shrinkage. For example, RTLS tags can alert security to commonly stolen items such as wheelchairs that pass beyond a perimeter threshold outside the building. Large hospitals are notorious for being equipment black holes. Was the device really stolen or was it just misplaced? Some IT departments have a staggering return on investment projections for this area.

Condition Monitoring   Temperature and humidity monitoring is required in many hospital departments, including the pharmacy and surgical departments. RTLS can not only track the temperature historically it can identify any variances that may jeopardize the product. For example in the case of organ transplants, where the temperature of an organ may literally mean life or death, the system can also provide the location as it moves through the hospital facility.

Patient and Clinician Safety   The hospital environment can be quite hostile at times. The safety use cases are very attractive to both hospital staff and management. Patients under the influence of a chemical substance or in a traumatic incident can become very combative

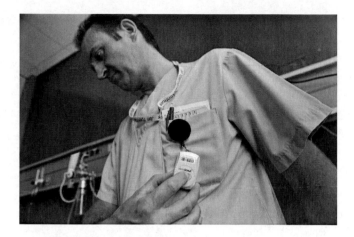

Figure 8.5 Clinician active tag. (Photo courtesy of Stanley Healthcare.)

and cause harm to clinicians or patients. By the push of a button on an RTLS badge, security officers can receive a distress alert providing the clinician's name and location, thus increasing response time (Figure 8.5). Some manufacturers have built in a distress feature into VoWi-Fi phones. This provides the same alert as the badge but also places a call to the security officer or desk on a speaker phone. A security officer then has multiple sources of information, including who triggered the distress, the exact location, and possibly even an audio feed of what is transpiring.

Patient tracking can help keep clinicians and patients safe. Such a system can alert staff in the event a high-flight risk patient moves beyond an allowed area. If a patient crosses a "choke point," an area they are not permitted beyond, an intercom system can alert everyone on the floor. This type of monitoring can save significant money by eliminating the need for a clinician to watch a high-risk patient.

Infection Control Hospital systems are more than ever under financial pressure to prevent spreading infection. A provision of the Patient Protection and Affordable Care Act, , Section 3025, added Section 1886(q) to the Social Security Act, which created the Hospital Readmissions Reduction Program. This legislation requires The Centers for Medicare and Medicaid Services to reduce payments to Inpatient Prospective Payment System hospitals with excess readmissions. As a result many healthcare organizations are heavily enforcing policies that will help reduce infection, thus maximizing reimbursements. One of these policies is hand wash compliance tracking. RTLS is able to track hand-washing compliance. Some solutions correlate clinician's badge tags with a smart dispenser that generates an electronically saved log to track compliance. As Orwellian as this may seem, the fact is an audit log of compliance for hand washing may positively impact patient outcome and hospital revenue.

Another use case pertains to staff quarantine. The outbreak of a serious airborne illness will require staff to be quarantined to minimize the spread of infection. With proper location tracking it may be possible to quarantine only staff members that were in the contaminated area, resulting in significantly fewer sidelined staff.

Workflow   Combining business intelligence with context awareness has the potential to make huge advancements in hospital workflow. Many hospitals are using RTLS to monitor patient experience. Using time stamping with location services can produce actionable alerts such as that the patient in an emergency room has been waiting an excessive amount of time. Tracking the time spent for each stage of the emergency department process from check-in to discharge can be used for process improvement.

RTLS Issues

Privacy Concerns   There is no doubt that the utilization of RFID and tracking can be viewed as "big brother" for some individuals. A properly designed system integrated with business intelligence can virtually track every minute of your day at work. One could easily be uncomfortable with the ability to track how long one spends in the bathroom each day. A number of labor unions call RTLS tracking an invasion of privacy, and are against the tracking of employees. On the other hand, the safety and other benefits described above may far outweigh the privacy concerns for both employees and patients.

Challenges with Accuracy   In almost all hospitals a tag will not have direct line of sight to the receiver except in the hallways. This is one of the most challenging issues to overcome as many RTLS systems require a direct and clear line of site to accurately determine the precise location of the tag. As mentioned above, fingerprinting may be a way to overcome the visibility issue. If the locations contain distinct measurement fingerprints, line of sight is not necessarily needed. In most cases this will require a number of receivers to facilitate a good baseline.

Maintenance and Costs   It is important to understand what is involved in maintaining an RTLS system. An RTLS system is comprised of many critical components that, like any hospital IT system, should be maintained, optimized, monitored, and backed up. Costs vary significantly from vendor to vendor. An understanding of the total cost of ownership of each vendor should be calculated. Many vendors also provide return on investment figures and even calculators to justify the investment in RTLS technology. Be sure to factor in the long-term costs of replacing batteries in both tags and exciters. In addition, most solutions entail some form of server to host a central location engine, and the hardware and ongoing maintenance of these servers should be clearly understood. Many vendors have a per tag cost model for support. If choosing a Wi-Fi based system the need to add additional access points could add significant costs that need to be evaluated carefully. Highly trained staff will be required to keep a sophisticated RTLS system calibrated. Be sure to assess if this expertise is included in the vendor's service package or whether you will need to factor additional costs to add or augment existing staff. Understand and consider maintenance costs that may impact your operational budget.

The integration and advancement of RTLS is certain to address today's challenges of tracking assets of all types. The possibilities of RTLS may be limited only to the imagination. The many uses for this technology have prompted significant investment in RTLS in many health organizations. The U.S. Department of Veterans Affairs will invest $543 million in room-based Wi-Fi RTLS across 152 medical centers in the next five years (http://www.nextgov.com/technology-news/2012/01/va-not-budging-on-plans-to-electronically-track-hospital-staff/50455/). The numerous technologies, including ZigBee, Wi-Fi, IR, and ultrasound, can all become quite confusing for the hospital. We recommend finding the right RTLS vendor who provides a complete hospital solution that is scalable and cost effective. This will include technology that meets the location accuracy needs of your organization, a software suite that aggregates location and management data into a single system for ease of clinician use, and integrates the use of business intelligence technology to help actualize cost savings and efficiencies from your RTLS investment.

9

THE WIRELESS PROJECT
MANAGEMENT PROCESS

Since its inception in 1984, the Project Management Institute (PMI) has made strides in formally defining and documenting the project management process and industry standards. One of the key books that the PMI helped craft and publish is *The Guide to the Project Management Body of Knowledge* (also known as the PMBOK). This serves as the foundation for preparations to become a project management professional (PMP). In 2013 there are 370,000 PMPs globally, and the number is steadily growing. The intent of this chapter is not to go into the project management processes in depth, but to understand how these are pertinent and relevant to wireless projects in healthcare.

The high-level processes that are part of a typical project are:

- Initiating
- Planning
- Executing
- Monitoring and controlling
- Closing

All of these play a major role in Wi-Fi deployment in healthcare, and will be addressed in the context of real-world experiences and challenges.

Project initiation is one of the most important processes in the project lifecycle. In the case of installing Wi-Fi in a hospital, this process can be broken down into developing a project charter and clearly defining the scope of the project. If we dive deeper into these concepts, the project charter entails obtaining approval from the health system or hospital leadership to pursue the project, and documenting a formal statement of work. A Wi-Fi deployment in a hospital can have a wide range of stakeholders, and all of these should be engaged during project initiation. Some of the typical departments that are key stakeholders are facilities, clinical engineering, marketing, IT, and the clinical community. It is crucial to identify a project champion to help keep the project momentum moving forward. The champion would ideally be an executive in the organization who has the influence and power to overcome hurdles as they arise.

Once the stakeholders are fully engaged, a project scope statement can be developed. Project delays can result if this document is not crafted carefully. Although facilities and marketing departments seem like minor stakeholders, they can contribute significantly to scope creep. For example, costs can increase quickly if marketing has a requirement to implement a multilingual guest portal on the wireless guest network, or the facilities department has specific aesthetic requirements for installing wireless equipment. Facilities can cause delays if it is not on boards with the way that the cabling vendor is installing cable, penetrating the facility walls or if fire-stopping is not completed correctly. Examples of typical requirements from the various stakeholders are shown in Figure 9.1.

The list of requirements can be expanded to include hand-washing tracking, temperature monitoring, QoS provisioning, etc. The more key stakeholders identified, the better the overall charter and scope, and the less likelihood of ongoing scope creep.

Department	Sample Requirements
Information Technology	Wireless network must have the capability to be centrally managed, and configured, and have the capability for policy enforcement.
Clinical Engineering	Wireless network must be able to support a variety of wireless medical devices ranging from legacy devices that can only support WPA2 PSK, to devices that can use EAP TLS. In addition wall to wall coverage including stairwells and elevator shafts is required.
Clinicians	Wireless network must be wall to wall and capable of supporting wireless employee access, guest access, medical device access, and voice access. Wireless network needs to support the use of smartphones, and support BYOD
Facilities	Wireless enclosures need to be aesthetically pleasant, must be hidden from patient view. Access points cannot be installed patient rooms, and the system needs to support RTLS use cases.

Figure 9.1 Departmental requirements.

Key stakeholders that are often left out of initial conversations are the departments that support end-user devices like the voice team, desk side support, the helpdesk, and mobile device management team. It is advised that they have a seat at the table given that the majority of wireless issues in the enterprise are related to end-user device configuration.

One pitfall to be aware of while crafting the scope is to clearly understand the requirement for wall-to-wall access. Often, organizations will attempt to save cost by proposing isolated pockets of Wi-Fi coverage, for example, in their larger conference rooms, but the reality is that these projects tend to become wall-to-wall deployments over time. Rather than augmenting a small Wi-Fi footprint, or incrementally growing the network, it is worthwhile to survey and design the network with the assumption that it will eventually grow to provide wall-to-wall coverage.

The planning process is typically the most labor intensive and time consuming. It is comprised of several dozen subprocesses, but the most crucial are outlined below.

Refining the Scope

As stakeholders become more engaged in the planning process, the scope of the project may change to accommodate requirements that were not clearly understood, or may have been overlooked. A good example is the head of the emergency department indicating that the department needs video remote interpretation services or telerounding and that the wireless architecture should take this into account. Another example is the requirement for Wi-Fi guest access in the cafeteria areas and outdoor picnic areas. There are several use cases for providing Wi-Fi access in the ambulance bays to facilitate the transfer of information from the ambulance to the hospital or clinic network.

Scheduling and Developing Milestones

As technical and nontechnical resources begin to work with the project manager and brainstorm the series of events that need to take place to complete the project, a work breakdown structure (WBS) is developed. This can be refined to include a high-level schedule and identify major and minor milestones. Some key milestones for a typical wireless project may be completing a predictive wireless survey, wrapping up onsite surveys, completing requests for proposals (RFPs) for cabling, completing the cabling, allocating VLANs and IP scopes as needed, installing access points, validating coverage, etc. In hospital environments, there are very few outage windows for maintenance, so these will need to be scheduled with the various departments ahead of time. In addition, due to dust containment concerns, tents are required to do any type of work above drop ceilings, so these need to be scheduled and allocated far in advance with the cabling vendor of choice.

Developing a Budget

Ensuring that the high-level plan is understood is directly correlated to the success of the project. This can include human resources as well as buying the necessary tools to complete the work. Wireless engineers and architects with the right level of certifications and experience are highly skilled and specialized, so it is key to budget for these resources in the project plan. Site survey tools can be fairly expensive as well. It is not out of the ordinary to spend upwards of $20,000 on a survey kit with all of the associated tools. In addition to these costs, it's important to clearly anticipate, understand, and break down the ongoing operational costs as well as any capital expenditure required. The lifecycle of the wireless hardware needs to be clearly understood, and the annual maintenance and support fees should be part of the overall budget. On occasion, training may be required for technical staff and this should also be part of the budgeting process.

In some cases, government or private grants may be available for use to augment available technologies in a healthcare institution. Special care needs to be taken with this type of funding in regards to payment terms, as well as potential time constraints. The conditions of the grant need to be clearly understood to ensure that project funds can be available in a timely matter.

Quality Assurance

As the project progresses, the quality plan becomes a roadmap for ensuring that the project is on track, and that it will meet any government/regulatory standards as well as standards identified by key stakeholders. These can include HIPAA and HiTECH, among others.

Communication Strategy

This is often overlooked, but can mean the difference between a successful and an unsuccessful project. The information and communication needs of the project stakeholders must be clearly understood. One real-world experience that shed light on the importance of clear communication was a series of interactions with a facilities department during a large-scale deployment several years ago. The project team assumed that facilities would be onboard with installing wireless access points with internal antennas on the wall around an 800-bed hospital, and that it would be alright to locate wireless access points in patient rooms. As the surveys were completed and the cabling vendor was getting ready to start the cabling, the facilities department indicated that the APs needed to be out of sight and that the team was not allowed to place APs in patient rooms. The result was a series of APs installed in the hallways and enclosures that added thousands of dollars to the project budget. If facilities had been engaged as part of a clear communicating strategy, these hurdles could have been avoided.

Another example has to do with Wi-Fi guest access. The guest splash page was designed with ease of use in mind, so it is a consistent page across the health system. The marketing department wanted to utilize custom splash pages at each hospital to highlight the main events taking place at the site and integrate multilingual support and a site map. The redesign can take several months to complete and cost several thousand dollars.

Risk Management

Project risks can impact timelines and budgets significantly if they are left unmanaged or not prioritized. Deploying Wi-Fi in new hospital construction highlights the importance of clearly understanding risks. The construction company working on building the hospital required that all ceiling-mounted AP enclosures be anchored to the ceiling foundation. Part of the challenge was identifying the exact

AP locations based solely on CAD files, because the actual walls were not up in certain cases. The risk of having to move an enclosure after the enclosures were anchored was identified and a longer service loop than usual was required.

It is difficult to project the amount of load on a wireless network prior to rolling it out. An assumption can be made that each user will have two to three wireless devices but in reality, you can run into a scenario where your network becomes a small ISP for guest users with thousands of concurrent users per day. This type of load can necessitate IP subnet redesigns and expansions, and can often result in having to validate the usability and performance of unique types of devices.

When a cabling vendor is the vendor of choice after an RFP that does not mean that they will perform quality work. It is important to outline the requirement to test each and every cable drop, and to take into account port capacity on the existing power over Ethernet (POE) capable switches and the need for new switches.

Change Management

Change management is typically mature within healthcare IT organizations that deal with operational support. The change management documentation needs to address technical as well as nontechnical changes that are part of the project. In addition every change needs to be compliant with the change management process, as well as Change Advisory Board (CAB) reviews. This serves as an audit trail and allows key stakeholders to be aware of any changes pertaining to the project.

Closure Criteria

No project can be successful if the criteria for success are not clearly defined. The closure criteria allow the team to report on the success or failure of the project. For example, if a wireless site survey is conducted to support data devices on 802.11b/g, there may be some issues supporting 802.11a voice handsets on the network.

As a way to illustrate the typical steps that need to occur during a wireless deployment, below is a detailed breakdown of the key milestones with an overview of the pros and cons. The plan below breaks down the steps for deploying Wi-Fi in a 500,000 square foot, three-storey state-of-the-art healthcare facility:

1. Identify Key Stakeholders and Set up a Kickoff Meeting

The point of clearly identifying stakeholders was addressed earlier in the chapter. Identifying and being able to see eye to eye with key stakeholders can help ensure that the project remains on track with as few disruptions as possible. Expectations can be clearly understood, and the success criteria can be identified and agreed upon. Although some typical examples were provided earlier, the example below may shed new light on this step.

During the build out of a new 200-bed hospital, it was indicated by clinical staff that Wi-Fi connectivity is a requirement in elevators and stairwells. The technical design team came up with several strategies to provide this type of coverage. These included putting an AP inside the elevator, installing APs outside the elevator on each floor, and installing wireless antennas at the top of the shaft pointing the semidirectional RF signal down the shaft. There were challenges with each of these approaches, but in an effort to provide the best quality coverage, the team decided to use a combination of APs outside the shaft and an AP in the elevator with signal pointed downwards. The stakeholder that was left out in this case is the elevator safety commission. When they learned about the plan, they were vehemently opposed to it, citing security and safety concerns. The majority of these were unfounded so after several months and hearings before the safety commission, the team was able to attain approval to move forward with the plan on a handful of elevator shafts. Several months later, with officer turnover, the commission revoked any additional growth. In this case, including the elevator commission as a key external stakeholder may have helped avoid some of the delays encountered.

2. Perform an RFI and RFP to Choose a Wireless Vendor

The RFP is a key process for ensuring that all appropriate technical and nontechnical aspects of wireless vendors are scrutinized prior to selecting which one to employ. This is an opportunity to gauge technical pros and cons of each product line in addition to gauging the financial viability of each vendor, and identifying a high-level architecture to use for comparison. Often, a handful of vendors can be eliminated based on poor, untimely. or no responses to the questions. All vendors should be subjected to an identical list of questions and provided with the same duration of time to respond, with a final response or additional clarification requirements. Once the list is narrowed down to two or three vendors, the process of elimination can become more challenging, and a scale to gauge the performance of each vendor will need to be developed. On occasion two or more vendors will have fairly equivalent product capabilities, and the decision boils down to cost as well as accessibility of the support organization.

3. Survey Network Closets for Port Capacity and POE Availability

A step that is often overlooked is ensuring that the closets in the facility have adequate room to accommodate the projected number of wireless access points and sensors. This includes ensuring that POE switches are installed and that they can handle the power load required from the model of APs chosen. If certain closets do not have room for expansion, this can translate to a requirement for a new network closet which can add some cost to the project.

4. Perform Predictive and Onsite Wireless Survey

Chapter 3 covers the site survey process in depth, but the steps below are a high-level outline of the process.

Step 1: walkthrough and questionnaire. An initial walkthrough of the facility must be planned. Some areas that need to be noted during this stage are:
- Whether door materials are wooden, metal, or composite.
- Behavior of the doors during emergencies.
- EMI (electromagnetic interference)-producing equipment that can cause interference.
- HVAC locations and layers above the ceiling, and the type of ceiling.

- Identify switch closets and how they map to the floor or building layout.
- Submit and review a site survey questionnaire with onsite IT professionals and the customer to ensure requirements are captured.

Step 2: acquire facility drawings. Up-to-date architectural drawings, including interior and exterior walls, doors, windows, elevator shafts, and stairwells, must be acquired for the facility being surveyed. 2000 or 2004 CAD format in DWG, DWF, DXF, and DWX extension are required. One drawing for each floor should be provided encompassing the entirety of the facility.

Step 3: import. CAD drawings must be imported and processed to build a model. This may be as simple as a multistory building or as complicated as several interconnected multistory buildings especially if campus mobility is required.

Step 4: attenuation measurements. A WLAN team of engineers must be assembled that will remain constant throughout the life of the project. Familiarization is a key element. Scheduled access to key locations will need to be determined and coordinated. One or two members of this team will take sample attenuation values for building materials and mark up points of interest on the floor plan printouts.

Step 5: spectrum analysis. Another member of the WLAN team will need to perform spectrum analysis sampling and measuring for noise levels. Part of the process is to understand what other entities are living in the 2.4- and 5.0-GHz spectrum. This will result in noise and interference on the WLAN so it is best to understand what is out there and determine how to address it. This is part of the predeployment strategy.

Step 6: understand customer requirements. Once the questionnaire has been completed and returned, an understanding of the type of WLAN that is going to be deployed and how the bands are going to be used (voice, data, voice and data, video) is crucial. Understanding the dependencies with the wired infrastructure is key during this step as well.

Step 7: generate predictions and review with the customer for their input.

Step 8: take measurements and validate predictions. Predictions should be validated onsite by using a method called AP on a Stick. Measurements are taken and used to calibrate the predicted values with actual readings. Adjustments to AP positioning can be made at this stage. It is critical that the physical structure of the building including all of the walls reflects the building state that the wireless LAN will be deployed in. Any areas undergoing renovation or construction will require a revisit and a resurvey.

Step 9: generate a bill of materials (BOM). Now that the predictions are finalized, and the total number of access points required for the project are determined, a BOM can be generated.

Step 10: final site survey report with detailed BOM. This is where all information to present to the customer is compiled, which will include the final report, all CAD drawings with access point and sensor predictions.

Step 11: certification. Once the WLAN is operational (access points and controllers installed), engineers are required to return to the facility with survey tools to take post-survey measurements, identify problem areas to correct, and certify the WLAN is functioning per design.

5. Develop Detailed Physical and Logical Architecture

After the site surveys are completed and the BOM of wireless APs, sensors, controllers, and management platforms is created, the architecture documentation can be created. These should include the physical locations of all wireless devices, including a layout of the network closets and racks. Often the wireless controllers are a 1 U footprint, but they can range up to 3 U. The physical location documentation is what the cabling vendors rely on to run cabling for the equipment. Some cabling vendors prefer to have a dedicated engineer pointing out exact installation location as well as antenna orientation, while others are OK with a colorful sticker on the ceiling illustrating AP location.

In addition to the physical design, the architecture phase is when conversations around design need to occur with the wired architecture teams as well as the security team.

The IP/VLAN allocation for the wireless equipment needs to be ironed out, and the end-user subnets should be provisioned to allow for 30 to 50 percent growth in the following 6 months. The design package should be clearly documented to avoid any confusion during execution. Any integration with existing systems should be documented in the architecture. This includes any DMZ/firewall related architecture being used for wireless guest service as well as integration with core or distribution switches, and existing voice gateways. The design should also outline and list the end-user devices that are supported.

6. Develop a Survey Report and Create a Cabling Bid Package

The survey report needs to clearly define assumptions, constraints, and risks associated with the requirements and the design. If a hospital indicates that video over Wi-Fi support is required, but does not want to focus on a QoS strategy around this, the risk needs to be identified. Some customers assume that Wi-Fi coverage in conference rooms is sufficient, but it has been shown over and over again that wireless clients outside the conference rooms will attempt to log onto the wireless network if they have a usable signal in their area. The coverage criteria and RF footprint needs to be clearly defined in the site survey with focus on the types of devices, the frequency ranges, and applications that the survey is designed to address. It should be clearly documented if the survey was conducted to support voice on the 802.11a/n or 802.11b/g/n bands.

The document has inputs from the survey tool, which illustrate coverage patterns as well as the distance to the nearest closet. Based on this information, the number of wired drops and available ports per closet can be identified. These can then be compared to what is available and augmentations can be planned accordingly as needed. In addition POE load on the switches can be determined. The cabling bid package can be developed based on the site survey documentation. It should outline the number of drops required as well as cable testing, fire stopping, and any Joint Commission on Accreditation of Healthcare Organizations (JCAHO) requirements around using tents to perform installations above the drop-ceiling grid for dust containment. In addition, cable labeling requirements as well as work schedule expectations, aesthetic expectations, and who needs to purchase patch cables, all need to be clearly outlined.

7. Order Hardware and Consider Lead Times on the Project Plan

In parallel with the bid for the cabling, the rest of the BOM can be ordered. This is typically comprised of wireless access points, sensors, IPS/IDS management systems, and wireless controllers. It's a good idea to order an additional 5 percent more access points than planned for use as onsite spare units in case of equipment that doesn't work when it is delivered. The lead time on this type of equipment to be delivered can be several months and this needs to be taken into account in the project plan. In addition, ongoing maintenance contracts need to be clearly understood. Depending on the wireless guest access solution selected, some hardware may need to be added to accommodate this as well. This can include firewalls and other dedicated appliances. Some wireless guest access solutions rely on an external vendor to provide splash pages as well as content filtering. This would be the time to finalize the statement of work and understand the cost model for these services.

8. Identify Third-Party Training Requirements

The wireless team provisioned for deploying the solution may need specific training to work with the product at hand, or to become more proficient with the technology at a high level. It's not unusual to send a handful of engineers to Airmagnet training or certified wireless network professional (CWNP) training so they can acquire the skill sets required to support the network once it is in place.

9. Stage Hardware

When the hardware arrives, the first step is to document the inventory, including serial numbers and MAC addresses, and to ensure that the hardware is in a secure location (Figure 9.2). Once this is completed, a unique asset tag will need to be attached to each piece of hardware. The nomenclature used in the site survey document should be attached to a highly visible area on each access point and

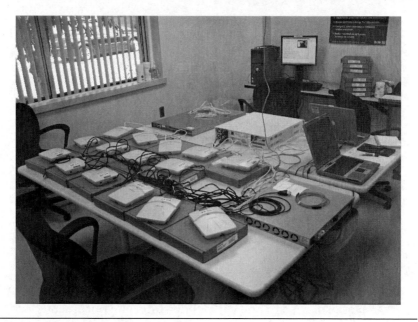

Figure 9.2 Hardware staging.

sensor. Following the naming process, each unit will need to be powered on to ensure that it works. The wireless controllers should be powered on for several days to bake in and ensure that they don't have any manufacturing faults. During this period, the access points can be staged on the wireless controllers to ensure that they have the appropriate configuration once they are online at their final destination. This is to allow the process to be plug and play once these are in the hands of the cabling vendor. In order to do this, a DHCP server will be required. A formal check off and handoff checklist can be used to indicate the status of each AP including firmware revision and labeling. This can then be used as a formal handoff document to the cabling vendor to alleviate any liability if hardware is lost somewhere along the way.

10. Oversee Installation and Turn-up of Wireless Network
Using a Standard Change Management Process

The physical installation of the controllers and access points may seem like a trivial step, but this is far from true. The change management process to obtain approval for the installation and the required technical reviews can take a couple of weeks. Standing up a new wireless

Figure 9.3   Example of poor cable installation.

network should have minimal disruption on the network, but ensuring the entire right configuration is in place on the DNS, DHCP, and RADIUS servers is key to a smooth deployment.

The placement of the wireless controllers and the IPS/IDS management platform need to be clearly understood and documented with the assistance of the data center raised floor manager. The cabling must be run in accordance with best practices (Figure 9.3). Figure 9.3 illustrates an example of a poorly installed Wi-Fi access point.

The access point installation needs to be treated as an end-to-end endeavor, and the cable testing should span from the switch through the patch panel to the AP. The cabling vendor should be fastening ceiling enclosures or brackets appropriately, as well as labeling each cable and patch panel, and coordinating with wired engineering to label each port on the switch. Wireless engineers can replicate these labels on the controller to ensure that finding where an AP is physically connected will take a minimal amount of effort.

Areas with special aesthetic considerations need to be pointed out to the cabling vendor. It is best to have someone work with them to show them how to address the installations in these areas. Sometimes these requirements are unrealistic: For example, a high-level executive with a conference room that has mahogany walls surrounded by thick concrete walls requesting wireless coverage for a voice device without allowing the team to install any APs in the room.

Certain high-profile areas may require custom antennas which will need to be certified for use with the product, which can take several weeks to do, so the general preference is to utilize antennas that have already been certified for use.

11. Ensure that All Hardware Is Set up on the Enterprise Monitoring System

The majority of wireless controllers can be added to management systems for SNMP reporting and management. This can be configured to initiate an alarm in the event of a wireless access point or controller going offline or experiencing some sort of issue. The granularity of the alerts can be customized to meet team requirements. For example, it may be sufficient to receive an alert when a controller is offline and its access points have failed over to the secondary controller. Depending on the environment it may not be necessary to receive an urgent alert every time a single access point goes offline.

12. Validate Channel and Power Plan

Many wireless platforms support automatically provisioning channels and power.

It may be a good idea to stand the initial system up using the auto channel/power feature, but in order to ensure that voice and other sensitive applications can perform up to a certain standard, it is usually a requirement to manually set the channels and power. This will allow for validating minimal co-channel interference and an adequate RF footprint.

13. Conduct Post–Implementation Survey and Make Modifications as Needed

The post-implementation survey is crucial for validating the overall health of the wireless network and channel plan. It should be conducted using a protocol analyzer as well as a standard laptop client. This step will provide the high level of confidence required to guarantee a certain level of performance. It should be kept in mind that access point augmentation may be required if there are areas found with inadequate coverage.

14. Perform UAT (Unit Acceptance Testing) Using
Various Form Factors of End-User Devices

The performance of one device on the wireless network is not representative of the health of the entire network. Each device type that is in the scope of the project should be tested thoroughly to ensure that any driver updates, or customizations are captured and documented. Every type of device will have a unique receive sensitivity, and roaming algorithms.

15. Send a Series of Communications Outlining Offerings with Instructions

After all of the functionality testing is wrapped up, and the various device form factors are tested, a series of communications will need to be sent out to end users to ensure that they are aware of the offering, and the requirements to access it.

The communications to employees can be sent out using standard e-mail distribution, but marketing may want to develop unique content for broadcasting the availability of the wireless guest network. Some departments have gone as far as integrating a link to the patient portal on an EMR system to the link. Others have developed stickers that are posted around the facility announcing the offering and indicating the support phone number.

16. Develop Helpdesk Knowledge Base for Common Troubleshooting

Some organizations have dedicated tiered help desks, while others host or use customer-support call centers. In either case, the staff can be trained through scripts of routine questions to decrease mean time to resolution (MTTR), as well as address issues quickly. Upwards of 90 percent of wireless issues that arise are client related. Half of these can be eliminated on the first call; for example, wireless guests having an issue accessing complimentary Internet that requires end users to launch a browser and accept terms and conditions. The helpdesk can walk these users through the troubleshooting process rather than engage higher levels of support.

17. Create a Runbook

The runbook is a design guide as well as an ongoing operational guide for team members. It is intended to be a document that can be provided to a new hire that contains all key relevant information to come up to speed on the design and the architecture of the network. This document can also outline ongoing maintenance procedures and expectations, including system firmware upgrades, support expectations, and road map. The runbook needs to be maintained on a shared drive, and should be routinely reviewed to ensure that it is kept up to date.

18. Handoff Support to Ongoing Operations Team

The formal handoff from the design and deployment team to the ongoing operational support team is often a step that is overlooked. This is because in some organizations these two teams are one and the same. Hospitals have little to no downtime, so ongoing support expectations and change windows should be agreed upon. Wireless controllers are often directly connected to the core or distribution switches which have strict outage windows. The service level agreements (SLAs) and service level objectives (SLOs) need to be clearly understood by the ongoing operations teams to ensure that they can continue to meet or exceed customer support expectations.

19. Ensure that a Process is in Place for Onboarding and Certifying Wireless Devices

The IEEE standards are constantly evolving, and the wireless architecture lifecycle is heavily dependent on these. On average a Wi-Fi network will need to be updated every 3 to 5 years to ensure that it is providing the latest and best technologies.

Wireless clients have a tremendous impact on the shared wireless bandwidth, so it is paramount that every type of wireless client being introduced into a hospital is thoroughly tested and evaluated for its potential impact on the network.

The process as illustrated in Figure 9.4 shows that this evaluation allows optimization to the client drivers as well as understanding of the potential impact of the device and the risks associated with adding it onto the network.

By including all of the key stakeholders in the onboarding and testing process, one can be sure that a device being evaluated is thoroughly scrutinized. In addition, the support model should be clearly understood prior to procuring the device and deploying it.

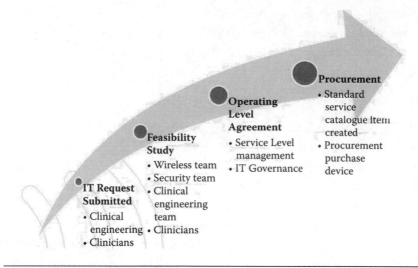

Figure 9.4   Example of an onboarding process.

10

SUPPORT CONSIDERATIONS AND LIFECYCLE

Supporting the complexities of the hospital wireless environment requires the right staff, the right knowledge, with the right tool set. The explosion of Wi-Fi-enabled clients will continue to grow increasing demand for advanced RF services. This chapter discusses critical elements needed to support a robust hospital Wi-Fi network.

Tool Set

A proper wireless toolset is critical to support the wireless infrastructure. Wireless technology has evolved rapidly over the last decade. In 1999, the 802.11b standard provided 11-Mb/s data rates. In 2007, 802.11n provided up to 450 Mb/s. In 2013, the 802.11ac draft has data rates reaching 1 Gb/s. As this evolution continues the need for cutting-edge tools is a must. Several different types of tools are required to regularly support and maintain the wireless environment. Appropriate budgeting for software and maintenance is something that cannot be overlooked. Many manufacturers require annual maintenance contracts to provide product support and upgrades. The next section will highlight each major category of tools necessary to efficiently support the infrastructure.

Protocol Analyzer

Wireless troubleshooting, optimization, and auditing have significantly advanced as a result of software specifically designed to provide the data needed to identify and fix client-related problems. Protocol analyzers are one of the first tools used when troubleshooting a wireless problem. The ability to analyze access point power level, channels, co-channel interference, and other RF-related issues is essential.

Some analyzers have a highly desirable option to import a spreadsheet that turns what is usually a MAC address to the actual AP name. Identifying the device name without the need to reference a map can be a huge timesaver.

Voice Analyzer

As discussed in Chapter 7, providing voice services over Wi-Fi can be very challenging. The AirMagnet VoFI Analyzer provides a unique way to analyze and capture data specifically to address voice issues. Voice is especially sensitive to latency, roaming, and interference. A VoWi-Fi analyzer (Figure 10.1) provides the ability to target and follow a specific device to track performance metrics like roaming and Mean Opinion Score (MOS). The MOS is used to assess the quality of VoWi-Fi communications. The MOS indicates the perceived voice quality of a conversation, ranking the call quality as a number between 1 to 5, with a rating of 5 being the best quality and 1 being the worst. The ability to correlate data from the phone, Wi-Fi, and wired network makes this is an extremely powerful tool. Utilizing this voice-specific tool can be very helpful in identifying and fixing the complex voice problems that are sure to be a challenge on any VoWi-Fi deployment.

Figure 10.1   VoWi-Fi analyzer. (Photo courtesy of Fluke Networks.)

Spectrum Analyzer

An often-overlooked aspect of Wi-Fi support is regular spectrum analysis. Spectral analysis should be used at regular intervals beyond the initial survey process. The rapid advancement in technology in the hospital environment brings with it the potential to introduce spectral interference from many different devices. Spectrum analyzers are available in handheld dedicated units or as software applications with a purpose-built adapter (Figure 10.2). Both meet the need for regular spectral analysis. Many spectrum tools provide built-in device signatures; therefore, identifying interference mediums like phones, Bluetooth, and microwaves can be automatically classified and located for faster remediation. Spectrum analyzers provide readings on RF energy, duty cycle, spectrum density, and other key metrics to help identify interference.

Site Survey Software

Many survey software programs provide other useful aspects beyond the survey process. Continual RF validation, performance testing, and capacity planning are a few useful features. Most suites now provide modeling for upgrading infrastructure to the 802.11n and soon 802.11ac standards. Site survey software was discussed in Chapter 3.

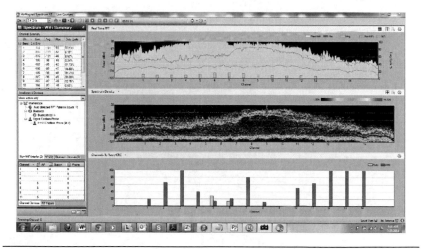

Figure 10.2   Spectrum Analyzer Pro. (Photo courtesy of Fluke Networks.)

Performance Software

Software that can analyze client-specific behavior on the WLAN can be very useful. Look for software that can provide Wi-Fi data without modifying the client supplicant. We use performance software to benchmark client device types, analyze networks before and after an upgrade, and assess end-user experience. Performance software can help understand roaming, throughput, and latency to pinpoint problem areas in the network for high-demand applications like video and voice. Figures 10.3 and 10.4 are graphics of Wave Deploy Pro software being used to illustrate good and bad client performances.

Using a floor plan and the site survey method of marking locations shows the actual performance of the device throughout the facility. The predominantly green color in Figure 10.3 represents an overall good user experience. This includes measurements of good TCP and UDP throughput.

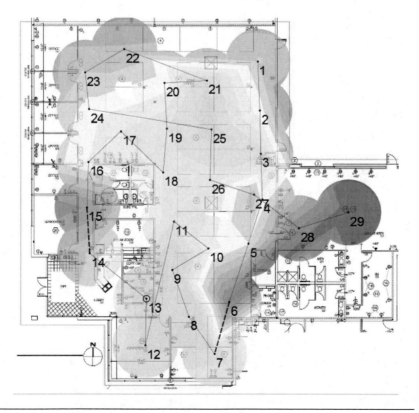

Figure 10.3   Good client experience. (Courtesy of Ixia Networks.)

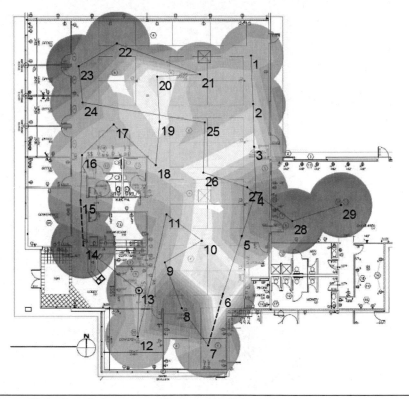

Figure 10.4   Poor client experience. (Courtesy of Ixia Networks.)

Figure 10.4 illustrates a bad client experience. The darker red colors exhibit failing TCP and UDP packets. The key to this is what the engineer has the ability to assess. This includes single and multiclient TCP and UDP, up and downstream throughput, MOS scores, roam times, jitter, and many other key network attributes.

Test traffic generator and performance analyzers are critical for the large enterprise. The ability to generate thousands of client sessions will provide capacity projections. Traffic can be accurately, repeatedly, and precisely created to assess performance in real-world conditions at scale. With client devices waiting to get on the network, with all of the WLAN variables, how does an administrator accurately predict not only the addition of a single client to the network but how multiple devices behave on the same infrastructure. It has become increasingly important to analyze not just one device but the many nomadic device interactions on the network. Additionally, the ability to benchmark each device via a certification and onboarding program

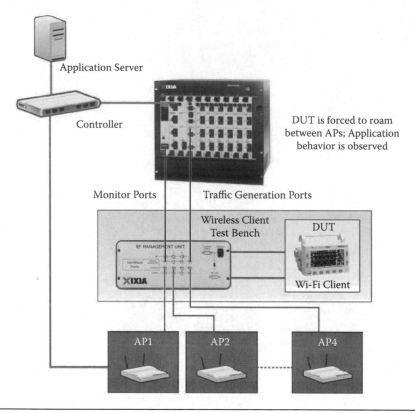

Figure 10.5   Ixia Waveclient. (Courtesy of Ixia Networks.)

can prevent issues prior to deployment by testing roaming, quality of service, and many other key metrics. Figure 10.5 depicts a standard configuration for Wi-Fi testing.

Packet Capturing

Packet capture and analysis is a required skill for the wireless engineer. The saying that "packets don't lie" is a very valid saying. It is important not to overlook the importance of utilizing this tool. Many packet capture programs are available for free. Packet capturing analysis to help resolve an issue seems to be increasingly underutilized. If packet capturing is not a skill on the WLAN team, take a look at the Certified Wireless Analysis Professional (CWNP) credentials. Comprehension of this material will turn any aspiring engineer into an analyzing genius.

Wireless Intrusion Prevention Systems (WIPS) An array of products provide options to monitor the RF environment for security purposes. This technology is often referred to as a sensory network. These sensors listen and classify RF traffic that may be harmful to the production environment. Most intrusion systems use signature-based attack mechanisms to classify threats on the network. Unauthorized RF devices or clients can disrupt your environment. There are two primary ways to implement WIPS. Both methods have advantages and disadvantages.

An overlay system is defined as installing dedicated devices used explicitly for security purposes. Overlay offloads processing power and resources the integrated approach has to account for. The overlay can also provide enhanced features to actively disrupt unauthorized devices or clients. Overlay may also be used to correlate wire side data with RF data. The overlay is more expensive as it requires the purchase and installation of dedicated devices.

The second method can be referred to as integrated. The access point switches into sensor mode when not under client load providing similar functionalities as the overlay system. Many wireless hardware manufacturers are beginning to offer dedicated radios to provide a hybrid option of the overlay and integrated systems.

Automatic rogue AP detection and triangulation is one of the most beneficial features of the WIPS. Figure 10.6 illustrates the location of a rogue AP. The location of the device is in the top right office.

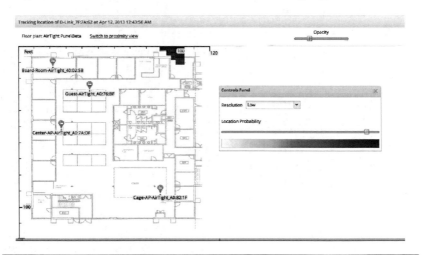

Figure 10.6 Airtight interface. (Courtesy of AirTight Networks.)

We use a sensor system that provides an alert in the event an unauthorized device is broadcasting in the network. This system checks the wired side of the network to identify threat severity. An unauthorized consumer access point plugged into the production network could provide access to the hospital network. These systems also provide compliance and security reporting. Regular reporting is essential to ensure your network is secure. Providing autogenerated monthly reports to the compliance team may be a requirement to "prove" the network is secure.

Wireless Network Management

Most wireless hardware vendors have created network management suites to help manage large infrastructures from a single console. These central platforms are very feature rich and help address many large-scale support challenges to providing centralized control and visibility in the enterprise. For example, the management console allows you to create central policies for role-based access controls. In the hospital this means you can group devices into logical containers to appropriately manage who, what, where, and how much clients have access to on the network. In the hospital you may want to provide different levels of access to doctors, residents, and guest users. Doctors may have precedence over these other groups by being in a higher QoS cue. Network management may also allow you to create self-registration for guest access and BYOD. With the thousands of devices this may involve, a central and automated way to manage access is essential.

Network visibility has become a much greater need for administrators. The ability for the management platform to identify client detail is a tremendously valuable tool. Figure 10.7 illustrates fingerprinting of a device. It provides information on where the device is, its IP address, operating system, authentication type, policy, and many other key metrics.

Staffing Considerations

Complex infrastructure requires unique skill sets. Wi-Fi engineering requires a sound understanding of not only the physics of the

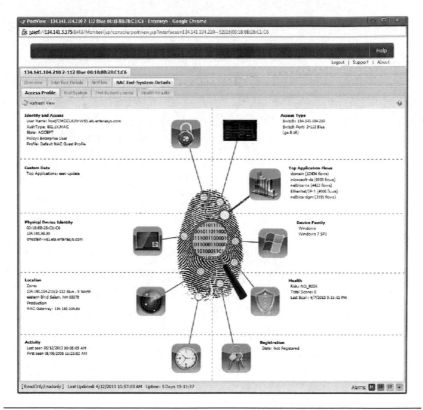

Figure 10.7   Fingerprinting. (Courtesy of Extreme Networks.)

propagation of radio waves but also mastery of network routing, switching, and security. The criticality and complexity of the Wi-Fi environment requires dedicated professional Wi-Fi engineers. Wi-Fi continues to provide increased throughput and performance. The notion of plugging in clients to the wired infrastructure will become significantly less feasible in the near future. This is evidenced by the number of new clients produced without a network port.

Traditional wired network engineering skills can provide a solid foundation for transition into wireless engineering. It is important to understand fundamental RF principles. We have encountered many untrained engineers who shared their misconceptions that wireless is something that could be thrown up in an environment and it just works. In the event this method doesn't solve the problem the misconception that simply installing more access points should fix it will lead to even more problems. This approach is unlikely to produce the robust wireless system needed in the hospital environment. Failure to employee skilled wireless engineers will cost your organization money and result in unhappy clinicians.

Formal Wi-Fi training and certification is highly recommended and available through a number of organizations. Engineers who hold formal certificates demonstrate to employers that they possess a certain level of knowledge about the job. There are three primary certification types: vendor neutral, software, and manufacture hardware.

Vendor Neutral Training One of the leading certification bodies in the wireless industry is Certified Wireless Network Professionals (CWNP). The CWNP organization offers a number of outstanding vendor-agnostic credentials ranging from very basic Wi-Fi fundamentals to expert-level design and analysis.

Entry-level training is the CWTS, or Certified Wireless Technology Specialist. The next level is CWNA, or Certified Wireless Network Administrator. We require all engineers to possess at least the CWNA certification as a requirement for employment. Three options are available at the professional level: security (CWSP), design (CWDP), and analysis (CWAP). The final level is CWNE, or Certified Wireless Network Expert. The highly coveted CWNE is one of the most prestigious Wi-Fi credentials in the industry (http://www.cwnp.com/certifications/).

The Sysadmin, Audit, Networking, and Security (SANS) Institute is one of the leading security training organizations for all security IT competencies. It offers the Global Information Assurance Certification (GIAC) programs. The Accessing and Auditing Wireless Networks (GAWN) is the wireless security offering. Candidates may choose to prepare for the exam by taking the SANS Training Course: SEC617: Wireless Ethical Hacking, Penetration Testing, and Defenses.

The Institute for Security and Open Methodologies began with the release of the OSSTMM, the *Open-Source Security Testing Methodology Manual*. It was a move to improve how security was tested and implemented. The wireless security expert credential is known as the OWSE. "The OWSE certification program is designed for those who want to learn more about the various ways to technically execute a comprehensive and professional wireless security audit within the internationally recognized Open-Source Security Testing Methodology Manual (OSSTMM) framework" (http://www.isecom.org/certification/owse.html).

Information Technology Infrastructure Library (ITIL) is the most widely accepted approach to IT service management. ITIL provides a cohesive set of best practices, drawn from the public and private sectors internationally. "ITIL advocates that IT services must be aligned to the needs of the business and underpin the core business processes. It provides guidance to organizations on how to use IT as a tool to facilitate business change, transformation and growth" (http://www.itil-officialsite.com/).

Certified Professional in Healthcare Information & Management Systems (CPHIMS) is a professional certification program specifically for healthcare information management professionals. "Passing the CPHIMS examination demonstrates mastery of a well-defined body of knowledge considered important to competent practice in today's healthcare information and management systems field. You will know that you have met the highest standards of practice and are among the elite in a critical field of healthcare management" (http://www.himss.org/health-it-certification/cphims?navItemNumber=13647).

Software Tool Training   Many wireless tool manufacturers also provide specific training in how to most effectively utilize their products. The latest software tools are very feature rich and can be best utilized with specific training. Ask your software vendor what training options are available with purchase.

Wireless Manufacturer Training   Hardware vendors all provide specific training and certification tracks. With the diversity of hardware vendor options, learning the "secret sauce" of the chosen vendor is a necessary evil. If your organization is making an investment with a vendor be sure to leverage training as part of the deal.

Ideally a well-rounded engineer will have a combination of vendor neutral, software, and manufacturer training. Participating in formalized training and certification will provide the knowledge you need to design, implement, and support the hospital-grade WLAN. Organizations should plan on annual training budgets to keep their staff up to date. It is an investment that will continue to yield dividends in the form of a healthy, robust network.

Wireless Runbook

The wireless runbook is a repository that documents all of the individual components of the wireless network. We refer to this document as the instruction manual for the entire wireless network. This document is very useful for both management and the technical staff for a number of reasons. The runbook includes details of policies, procedures, and architecture of the wireless network.

Policies

Network policies can be a very broad subject. For the purposes of this chapter we will highlight the most popular policies encountered in the hospital wireless space. Many hospital organizations have entire departments dedicated to acceptable use and security policies. The most important aspect of a successful policy is executive-level support. Without leadership support, enforcement is futile undermining the intent of the policy in the first place. The key policies pertinent to the WLAN are acceptable use and disaster recovery (DR).

Acceptable Use Who has access and from what device to information on the network is an important policy to determine. As Wi-Fi is a finite resource, meaning that the amount of spectral capacity does have a limit, should every device that has Wi-Fi capabilities be allowed on the network?

Part of the acceptable use policy contains the organization's response to demand for BYOD. Discussed in greater detail in Chapter 11, this policy should be carefully implemented. Sound strategies must be in place pertaining to the support and implementation of a BYOD policy. The most important aspect of BYOD is ensuring patient health information is secure.

Disaster Recovery A DR policy is used to recover both normal and critical business IT functions. This is essential in the hospital environment. Who has detailed information of the network to continue support in the event of a catastrophe? What if for some reason the technical staff is unavailable indefinitely? A sound DR policy documents the details to not only restore hardware and software functionality but also prepare for the element of manpower. A tornado may damage or destroy network equipment. An epidemic could disable key staff members. Document the procedures to recover hardware and software systems. A well-executed policy will be updated continually as new systems are brought online. A DR plan should be tested often to ensure the policies and procedures achieve the intended purpose. The runbook is a great place to store the DR policy for the Wi-Fi network.

Procedures Implementing proper Wi-Fi support encompasses many different procedures.

- Service desk escalation: The service desk has the difficult job to receive, attempt to resolve, and escalate problems to the appropriate support group as needed. This position is one of the most difficult jobs to do in the information field. A defined procedure for the priority and severity of the issue will help streamline support.
- After-hours support: Most hospital environments will require 24×7×365 support. Documenting the procedure for Wi-Fi support after normal workday hours should include escalation criteria, response, and resolution timelines from the business. A good approach will consider the criticality of the issue with the required response. Widespread outage impacting an emergency area will require immediate action, whereas a single client issue may warrant a less expedient response.
- Change control: Is a formal process used to ensure that changes to a production system are introduced in a controlled manner? The primary goals of a change control procedure include minimal disruption to services, reduction in back-out activities, and cost-effective utilization of resources involved in implementing change.

Architecture

Documentation of both the logical and physical architecture is included in the runbook. This should go beyond the standard network figures and should encompass all critical pieces of the Wi-Fi hardware and the supporting infrastructure. Document all hardware components including the supporting network infrastructure. This includes DNS, DHCP, and authentication mechanisms.

Systems Lifecycle

Routine Maintenance

A good maintenance program must focus on preventing system failures before they occur. Preventing failures means more productive clinicians. A way to reduce failures will involve designing and implementing procedures for recurring maintenance. The following should be performed on a regular basis:

- Perform a qualitative and qualitative analysis of the entire RF environment.
- Compare collected data to the original survey or last baseline recording.
- Comprehensive spectral interference analysis.
- Backup system validation and review.
- Security best practice review.
- Comprehensive equipment inventory audit.
- Performance and throughput testing.
- System diagnostics.
- Client device compliance audit.

Preventing system failures by implementing proactive maintenance increases efficiency and clinician downtime. In turn, Wi-Fi technology provides enhanced workflow efficiencies to help clinicians provide better patient care.

Technical Support

With the prolific use of Wi-Fi technology there is a growing requirement to provide end-user technical support. Technical support is simply defined as providing assistance to customers who encounter challenges or difficulties with technology and for the purposes of this book specifically, Wi-Fi technology. Support may be requested in a number of ways, including by telephone, e-mail, or a web portal where users can log a support request, sometimes referred to as an incident or trouble ticket, for assistance. Most hospital organizations have internal technical support to address technical problems.

Technical support is usually subdivided into multiple tiers in order to better serve a hospital customer base. A multitiered support system is utilized in order to provide efficient support services by dividing responsibilities by skill set. The critical success factor for implementing a successful multitiered support model is a clearly defined process for escalation. A support technician needs to understand all service level agreements/objectives (SLA/SLO). These metrics are determined by the business to provide a baseline for restoring service. SLAs usually define customer response and resolution timelines. These well-defined metrics should help the IT staff appropriately escalate to upper tiers. A common multitiered support model is a four-tiered technical support system but this will vary by organization.

Tier 1 Tier 1 is the first line of support. Tier 1 technicians receive the customer calls first and are often referred to as customer support representatives. The first job of a tier 1 specialist is to gather the customer's information and determine the customer's issue by analyzing the symptoms and figuring out the root of the problem. They perform triage, solve straightforward problems, often using a knowledge management tool, and pass unsolved issues on to tier 2 support. In some rare cases, they may escalate directly to tier 3 depending on the diagnosis of the issue. Support personnel at this level have a basic understanding of technology and may not always have the expertise required for solving complex issues. The goal for this tier is to resolve 70 to 80 percent of incidents on the first call without having to escalate the issue to a higher level. First-level support is usually carried out by representatives housed in a call center that facilitates support 24×7×365.

Tier 2 Technicians in this group are responsible for assisting tier 1 personnel in solving basic technical problems. This group is often referred to as technical support engineers. Their primary responsibility is to handle escalations from tier 1 and solve issues, often with the assistance of a known solutions database. Tier 2 engineers typically have direct access to higher-level tier 3 to assist in resolving a problem before escalation. Troubleshooting solutions may be performed by this group to help ensure the issues of a complex Wi-Fi network are solved by experienced and knowledgeable technicians. This usually includes onsite installations or replacements of wireless hardware, software troubleshooting and performance testing, and advanced RF troubleshooting. It is important that the engineer be well versed in the SLA/SLO metrics. If a ticket has been worked on for quite some time before receiving an escalated trouble ticket it may delineate the approach to find resolution. For example if the ticket is already close to a resolution time specified in the SLA, because of the time spent at tier 1 working with a customer, the engineer may need to escalate the issue quickly in order to meet the resolution deadline. This is a fundamental element required for meeting both the customer and business needs as it allows the technician to prioritize the troubleshooting process and properly manage time to resolution.

Tier 3   This group is responsible for the highest level of support. System administrators are responsible for resolving the most complex and difficult issues. System administrators are the most highly trained and skilled experts in the field. They are responsible for the highest level of troubleshooting and analysis. In some cases the third tier will be able to resolve the issue that one of the lower tiers may have missed. If this is a newly discovered problem this group is responsible for implementing the fix and providing supporting documentation to the lower tiers for future use. Tier 3 administrators will have the same responsibility the other lower tiers have, of assessing the time already spent with the customer so that the work is prioritized and time management is sufficiently utilized to meet SLAs.

Tier 4   This tier represents escalation beyond the hospital IT staff. This is generally the wireless hardware or software vendor. If the tier 3 administrators cannot resolve a problem, this is the last resort for problem remediation. It may not be uncommon for a vendor to have to produce a new code release to fix a newly discovered bug. Understand the support service levels that the vendor is accountable for. Often the support contract will have provisions for hardware replacement, response, and restoration. Continue to track incidents even when they are escalated to a vendor. Many organizations have specific provisions for the SLA tracking of vendor-escalated issues.

Infrastructure Code Upgrade

We have a saying that the quality of manufacture code running on your wireless equipment can equate to the quality of life of the support staff. Regular code upgrades are inevitable. Often there will be at least some downtime related to these activities. The critical nature of the hospital Wi-Fi network, supporting thousands of devices, makes it exceedingly difficult to schedule downtime. It is highly recommended to establish regular maintenance windows to increase the ease of performing necessary code upgrades. Maintenance windows should be carefully chosen to minimize impact. Late night changes are almost always going to have less impact than during the day. Saturday night may not be the best time for a facility with an emergency department as this can be a peak time. If the best time for a window is unknown ask a number of clinicians to find out. Don't be discouraged if many indicate there is never a good time.

Many vendors release upgrades at regular intervals. It may be helpful to establish a procedure to evaluate new code release. Choose to implement code that provides feature enhancements and bug fixes critical to the environment. Once code is chosen it should be rigorously tested in a lab environment to emulate the production system. If satisfactory results in the test lab are achieved, choose a pilot site that is low risk. Be sure to follow the change control process to help mitigate risk and increase the chances of a successful change. Once the code has been vetted it can be widely distributed with a high confidence level minimizing impact on patient care.

End-User Device Considerations

Client devices need to be kept up to date and should be refreshed on a regular basis. Outdated clients will have the least capable wireless functionality. Client devices that only support legacy data rates should be replaced as soon as possible. Legacy devices potentially slow the entire network. The age and types of devices may also need to be considered when upgrading a network to 802.11n or 802.11ac. Network enhancements will not be realized by outdated clients.

Keeping wireless network interface card (NIC) drivers up to date is also often overlooked. A network interface card driver is simple software. Like any software it is not immune to security exploits and bugs. Keeping the drivers up to date assures that wireless devices have the latest protection and features from the manufacturer. We have discovered a number of troublesome wireless connectivity issues directly relating to outdated or bad NIC drivers.

The increased security measures of HITech and HIPAA require the healthcare industry to be more accountable for patient health information (PHI). Securing the client device with anti-virus and malware protection is essential. Hard drive encryption may also be a requirement for devices storing PHI. A lost or stolen device containing patient data can cause serious financial, legal, and reputational implications to any hospital organization.

Lifecycle and Drivers for System Upgrades

Infrastructure Lifecycle   Wi-Fi infrastructure products have different lifecycles. As we showed in Chapter 1 on the history of Wi-Fi the evolution of wireless technology has occurred very quickly. In just over a decade Wi-Fi has changed from "nice to have" to an absolute necessity in the healthcare industry. This rapid evolution of product enhancements has yielded relatively short product lifecycles.

There are three primary classifications pertaining specifically to the final stages of a product lifecycle, including end of sale, end of support, and end of life.

End of sale means the manufacturer is no longer offering the product for sale. This may be for a variety of reasons, including that the product may be obsolete or components may be less available. Manufacturer support is typically still available after this stage.

End of support often means no other software or bug fixes will be produced and manufacturer support will cease.

End of life is when the product is completely retired. This usually means not only is the product not sold and supported but usually replacement parts are also not available. Product lifecycle from end of sale to end of life varies greatly by manufacturer. It is never recommended to have production equipment in the later stages of the lifecycle. It is a very risky proposition to have unsupported hardware in the hospital enterprise. Leadership should be made aware what the possible implications are of not appropriately upgrading an end-of-life system. To avoid risky situations budget appropriately. Creating a document that illustrates the current and future state of the infrastructure may help convince leadership of the need to provision an appropriate budget.

Client Device Lifecycle Managing IT equipment and product lifecycle is an important function of the IT department. A properly executed equipment lifecycle management program should reduce failures because technology equipment is replaced prior to failure. Ideally this program will reduce the total cost of equipment management over its lifetime. Managing wireless devices poses a number of specific challenges and complexities. Most IT departments have figured out how to manage common laptop and desktop hardware lifecycles.

The explosion of mobile and medical devices has introduced a number of new lifecycle variables. Many mobile devices are tied to a cellular service provider for a specific term. This will usually drive the lifecycle of the mobile device itself. It may save a lot of headaches to choose a device with a standardized operating system. A support structure that attempts to support every device on the market is an uphill battle for any organization. If the management demands a broad scope of devices it will be necessary to at least require a minimum acceptable code level.

Medical devices add a number of variables to consider regarding lifecycle. Many different business units have responsibility for these devices, including biomedical, IT, and many clinical departments. The medical device area is particularly challenging from a lifecycle perspective. Often the oldest remaining clients on the network are legacy 802.11b medical devices. These devices not only slow down the entire network, they also usually support less than desirable security options. Support models from medical device vendors may be cost prohibitive. Do not be surprised if code has never been replaced on these client types. This can be problematic as outdated code is often vulnerable to security exploits.

Mobile devices will pose additional lifecycle challenges. It is important to include emerging mobile device types into the corporate lifecycle policy. This will ultimately save money and increase clinician productivity, in turn improving patient care.

Technical support and lifecycle management is a critical success factor for managing any infrastructure. The need to appropriately budget and manage product lifecycle is a key business requirement. Highly trained support staff will help you save time and money. Keeping staff regularly trained will ensure the rapid evolution of Wi-Fi is appropriately managed at your healthcare facility.

11

EMERGING TRENDS AND TECHNOLOGIES

Trying to predict the future of Wi-Fi-enabled healthcare is an impossible feat, however, it's important for everyone providing IT services in a healthcare environment to remember that their responsibilities are in two unique environments, healthcare and technology. At the end of the day efficient, effective, and affordable patient care is the goal. Healthcare IT professionals need to keep abreast of the latest technology advancements in addition to the evolution of patient care delivery models, and the ever-changing operational strategies for healthcare organizations. One area that IT can bring to the table is insight and guidance on how emerging technologies can better support healthcare delivery. To quote the great Wayne Gretzky, *"A good hockey player plays where the puck is. A great hockey player plays where the puck is going to be."* Our responsibility is to improve an organization's ability to provide patient care and the patient's overall end-to-end experience.

Both healthcare and technology futurists like to pontificate on potential grand improvements in care delivery attributed to seismic shifts in the promised capabilities of new technologies. More often than not, the predictions fail to be realized. We call this the jetpacks and flying cars syndrome. After all, we are all still waiting ... and the astronomical costs of delivering care continue to rise at an alarming rate in the United States.

Rather than pull out a crystal ball or rely on the grand plans of technology vendors about the wireless utopia that awaits us all, a more pragmatic approach is to focus on more than 3 to 5 years out. Any farther and planning becomes a guessing game, past the next round of technology refresh activities, and of limited value to the immediate delivery of improved service to an organization's operations.

With all that said, here are some key up-and-coming technologies and healthcare trends that will have significant impact on the Wi-Fi-enabled healthcare landscape.

Demand for More Bandwidth and Denser Deployments

802.11n was thought by many to be the answer to growing bandwidth needs, but a number of trends are providing new challenges for WLAN infrastructures. These are all rooted in a scarce spectrum and an insatiable hunger for bandwidth from end-user devices and applications.

Device Density

As discussed earlier many Wi-Fi networks were initially designed both in terms of capacity and coverage for support of mobile PCs. One device per employee was considered a high design ratio. Smart phone and tablet adoption by employees and healthcare organizations continues to grow with no signs of slackening. "Today, an average user can easily require two, or even three connected Wi-Fie devices." With the number of clinicians who carry devices increasing, it's becoming common to find areas of insufficient RF coverage and device saturation. More coverage and denser coverage are becoming Wi-Fi design fundamentals in hospitals. Gartner predicts that 80 percent of corporate Wi-Fi networks will be obsolete by 2015 and companies deploying tablets will need 300 percent more Wi-Fi capacity to be effective. Additionally, the devices are requiring higher-quality networks in support of bandwidth-hungry applications such as imaging, or performance-sensitive real-time applications such as voice and video.

In addition to the smart phones and tablets carried by clinicians, healthcare organizations are also under pressure to deliver a high-performance guest network for patients and visitors, many of whom have the expectation that streaming video be supported.

Evolution of the Electronic Medical Record

Probably the single greatest justification for Wi-Fi system adoption in healthcare was in support of clinician access to the electronic medical record system (EMR). While this may have started with laptops and workstations on wheels, it is really the tip of the iceberg in an organization's long challenge ahead to adopt what Gartner calls the real-time healthcare system or real-time enterprise (RTE). The RTE is a management paradigm that uses new and existing information in a different way, accelerating individual processes to minimize response time and optimize quality. It's made up of numerous applications and infrastructure systems communicating together to provide clinicians and the organization with greater awareness in support of faster response and outcomes. In short, more devices and applications will be communicating. Examples include: people to machine (P2M) *clinical order entry systems*, machine to machine (M2M) *IV infusion pumps connected to the EMR*, and machine to people (M2P) *lab systems alerting clinicians*.

Mobile Voice and Video

Every PBX manufacturer supporting healthcare organizations today is highlighting the advanced features and capabilities for the mobile worker. What started as VoIP on the WLAN with dedicated devices (Ascom, Cisco, Spectralink, and Vocera) has now expanded to voice and video applications on mobile devices. Besides the enterprise communication applications on the network, use of consumer applications such as Apple FaceTime, Skype, and Google Hangouts used by patients and employees alike, will require more complex application-layer policy support and management on the infrastructure. End users will view performance of communications applications as the responsibility of IT regardless of whether they installed and administer the applications.

Guest Access

When we first started deploying guest access on WLANs years ago there was very little traffic. This had everything to do with the fact that most visitors to hospitals and clinics were not carrying laptop computers with them. Smart phones and tablets have changed that trend forever. Visitors now view guest access as a right not a privilege and, as discussed earlier, how well guest services are provided strongly affects a visitor's satisfaction with the healthcare organization. Growth of devices on guest networks and the bandwidth they are using shows no sign of slowing. We have witnessed the trend first hand with the influx of devices after the holidays and their demands on the network as patients choose to watch content and communicate on their personal devices rather than use the hospital-provided phone and television services.

Additionally organizations are starting to realize that the guest user experience is something that can be leveraged by the organization as departments begin using the captive portal pages for visitor education on a variety of services. Multilanguage support, patient EMR portal access, and other services will be a growing requirement to any guest services offering. The possibilities are endless when it comes to targeted marketing, but the cost is taking on the role of a small ISP, and ensuring that enough bandwidth is allocated for guest users.

Patient Engagement with Social Media

In a recent study covering mobile and social technologies in health care, it was noted that: "90 percent of people ages 18-24 said they would trust health information they found on social media channels. One in two adults uses their smartphone to look up health information." The study also found that more than one in four hospitals have a social media presence and that 60 percent of doctors believe social media improves the quality of care. Social interaction between patients and caregivers will continue to grow and the CIO, who is responsible for building the connected and social health care ecosystem, has the opportunity to be in the forefront of improving the patient-physician experience.

With the proliferation of mobile devices on the health care network, so does the use of social media as a means to communicate with patients. Today's medical students and young physicians have grown up in the social media universe, and they expect to use social media as part of their medical practices. Patients are looking to their network to share medical information and learn more about their conditions and treatments.

The health care CIO could be an innovator on the subject of improving patient care and enabling meaningful and timely communications between the physician and patient. Every day, millions of people sign up for Twitter, Facebook, and LinkedIn, and more and more are looking to newer sites such as Branches, Quora, and Pinterest. These people are all consumers of our nation's health care system and if there is no technology leadership driving adoption of social media in health care, the very people you are looking to engage will be largely ignored.

Today, many doctors are looking to use social media to not only communicate with patients, but also as a platform to keep patients well and informed. By keeping patients healthy, as opposed to just treating illness, doctors can lead the way for changing how our health care system is utilized and thus reducing the cost of care. However, if health care IT is not a leading voice for the very tools the physicians are using to reach their audience, there is no way to be a pioneer for enabling this communication. (Vala Afshar, CMO and Chief Customer Officer, Extreme Networks, and Tamera Rousseau-Vesta, Healthcare Solutions, Extreme Networks, "For Health Care CIOs, the Social Media Landscape Is a Ghost Town," Huffington Post, January 1, 2013.)

Device Consolidation

It's not uncommon for a clinician to have a cell phone, Wi-Fi phone, pager, and maybe even a barcode scanner. Companies such as Voalte are changing. They are offering specialized smart phones that can deliver via software application design with clinical workflow in mind (Figure 11.1).

"When I started in the healthcare industry, I was surprised to find systems being built on antiquated technologies. I knew we needed a better way for caregivers to communicate with each other and elevate the practice of caring for patients. In 2008, when Apple released its software development kit (SDK) for the iPhone, what had been a communication device suddenly became a powerful handheld computer. I founded Voalte to leverage the power of new mobile technologies and make a significant difference in the healthcare industry. Today, as consumer devices continue to advance, clinical applications and workflow are poised for advancement as well. At Voalte, our goal is to help hospitals embrace these new technologies and transfer their use into more efficient processes, happier clinical staff and better patient outcomes" (Trey Lauderdale, Founder and President, Voalte).

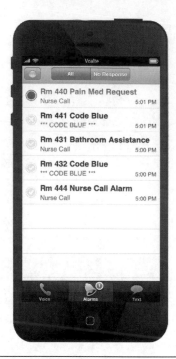

Figure 11.1   Voalte handset. (Courtesy of Voalte Inc.)

The use of pagers is expected to drop significantly as clinicians opt for software solutions that provide improved functionality on their existing smart phones. The challenge will be in designing Wi-Fi networks to match the RF coverage provided by pagers, which operate in the 929- to 930-MHz spectrum and are able to penetrate in many areas that are difficult to cover in buildings.

As discussed earlier, another challenge that many healthcare organizations are working to address is the wave of BYOD adoption. Regardless of whether the mobile device is owned by the organization or is an employee's personal device, the adoption of a mobile device management (MDM) solution is highly recommended. MDM solutions from leading vendors such as AirWatch and MobileIron provide IT departments the ability to remotely provision and manage mobile devices. Solution capabilities vary but essentially all allow centralized asset management, remote application of network configurations, and the enforcement of IT policies. Without an MDM platform mobile device support can quickly become not only heavily taxing on service desk resources but also a security risk. A new trend developing in these solutions is how both the MDM vendors and WLAN manufacturers

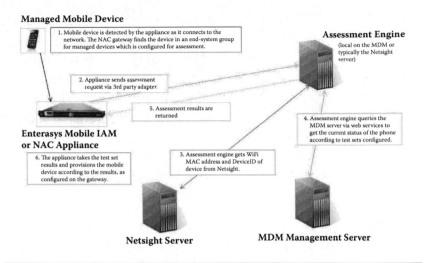

Managed Mobile Device

1. Mobile device is detected by the appliance as it connects to the network. The NAC gateway finds the device in an end-system group for managed devices which is configured for assessment.

Assessment Engine
(local on the MDM or typically the Netsight server)

2. Appliance sends assessment request via 3rd party adapter.

5. Assessment results are returned

**Enterasys Mobile IAM
or NAC Appliance**

6. The appliance takes the test set results and provisions the mobile device according to the results, as configured on the gateway.

4. Assessment engine queries the MDM server via web services to get the current status of the phone according to test sets configured.

3. Assessment engine gets WiFi MAC address and DeviceID of device from Netsight.

Netsight Server   **MDM Management Server**

Figure 11.2   Example of MDM integration. (Courtesy of Enterasys.)

have begun to collaborate on shared APIs so that functionality can be integrated between platforms. Figure 11.2 shows an example of this collaboration with Extreme Networks and AirWatch.

Shrinking Herds of Carts on Wheels (CoWs) and Workstations on Wheels (WoWs)

The largest obstacle today for adoption of tablets for computerized provider order entry (CPOE) has been due to the major EMR vendors' ability to port their applications designed for desktops to other form factors while maintaining the same usability. This will eventually change and when it does, the costly workstations on wheels will be replaced with large numbers of tablets and other handheld devices.

Key Emerging Technologies   Addressing the emerging trends will require a multifaceted approach to network solutioning and management at both the physical and application layers. New disruptive technologies to follow closely include:

- 802.11ac
- SDN and policy management
- IPv6
- HotSpot 2.0 802.11u

IEEE 802.11ac   802.11ac is the fifth generation of Wi-Fi standard, building on IEEE 802.11n in the 5-GHz spectrum. Its purpose is to improve data rates up to 1 gigabit per second, increase RF bandwidth utilization efficiency, and support denser access point deployments. For those wanting more details on the fundamentals of the standard the IEEE 802.11ac Wikipedia entry is a recommended site: http://en.wikipedia.org/wiki/IEEE_802.11ac.

To make things interesting for those looking forward to the expected bandwidth benefits of 802.11ac, the standard is being released in two waves:

- Wave 1Single User–Multiple Input Multiple Output (SU-MIMO): An AP has the ability to use multiple antennas to send data to a single client. The benefit is more bandwidth than what was available in 802.11n.
- Wave 2—Multi User–Multiple Input Multiple Output (MU-MIMO): An AP now has the ability to use multiple antennas to send data to multiple clients. This will be the closest Wi-Fi has ever gotten to replicating the full duplex capabilities of a wired Ethernet.

Realizing the benefits of 802.11ac will not be as simple as upgrading or replacing access points. Instead it will require coordination in the following areas:

- Infrastructure
- Client devices
- Design and planning

Infrastructure   Access points will be the first upgrade required. Depending on your client loads, bandwidth capacity in your wired switch network may create bottlenecks. Many AP vendors are planning to use 2-gigabit Ethernet ports to provide enough capacity back to the wired switch network. Troubleshooting AP-to-switch configurations just doubled in complexity with twice the switch ports and cable runs to now take into account.

Client Devices   When 802.11n was first released it took a number of years before supporting clients began to overtake legacy devices. It has taken even longer for first tablets and then smart phones to support 5-GHz 802.11a/n. Consumers also soon learned that the bandwidth capabilities of 802.11n were limited to how many antennas a client radio contained. The smaller the device the fewer the antennas is a good rule of thumb. With 802.11ac history will once again repeat itself in the order in which devices supporting 802.11ac make their way onto the market and the number of antennas they contain. Additionally, the first generation will most likely carry on the tradition in needing multiple firmware updates to address bugs. As 802.11ac devices begin to permeate the market, healthcare organizations with an 802.11 a/b/g/n network will start to see heavier client loads in the 802.11 a/n band and a steadily declining load on the 802.11 b/g/n network. This trend alone should result in improved performance in the congested 2.4-GHz space.

Design and Planning Before you can start replacing access points a thorough understanding of 802.11ac's properties as well as any optional features your vendor may have implemented need to be taken into consideration. 802.11n brought bonding two 20-MHz channels to 40-MHz for higher throughput into the equation. With that came challenges in channel planning and supporting legacy devices. The 802.11ac standard supports 80- and 160-MHz channels for higher throughput. This will lead to an even more complicated channel plan and challenges with supporting legacy devices.

Policy Management and Software Defined Networking (SDN)

In the early days of Wi-Fi WLANs were rather simplistic. The number of Wi-Fi devices on the market was limited and these were generally purpose built. Mission critical use cases were limited. VLANs have typically been the preferred way to segment and prioritize traffic. Laptops were assigned a VLAN, voice traversed on a voice VLAN, and 802.11n added the capability of supporting video on another. When medical devices became Wi-Fi enabled they typically were bundled into an additional VLAN. This design approach has two weaknesses that need to be taken into consideration.

The Rise of the Smart Phone

Most clinicians no longer want to carry multiple devices. What made smart phones so invaluable is the ability to add applications to infinitely increase the usefulness of the phone. Now instead of a dedicated Wi-Fi phone on a prioritized and isolated VLAN, one can carry a smart phone that is a VoIP phone, a video conferencing device, receive emergency pages, transmit medical diagnostics, etc. Each application may have its own unique quality of service (QoS) requirements. A given device can have a variety of applications ranging from mission critical applications to e-mail, personal entertainment, or wellness tracking applications. The device is no longer unique to a single type of application so a single VLAN assignment is no longer the best way to support these.

Application Performance and Security

One medical device on a VLAN is simple to isolate for performance and security. Today, it is not uncommon to have half a dozen different medical devices on a WLAN. Best practice for WLAN RF performance is to limit the number of SSIDs making a VLAN approach inefficient and impractical. It's also not uncommon to have medical devices that have limited support for the latest security standards, or require configuration management be performed by the manufacturer.

How does one address these new changes? Start with the implementation of policy management. Most WLAN solutions offer the ability to utilize policies to:

- Apply firewall rule sets to allow/restrict/deny traffic.
- Use different authentication and encryption settings per device.
- Set bandwidth limits.

With properly designed policy management and network access control an organization can provide role-based access control and bandwidth provisioning.

Moving beyond policy management is the growing interest in Software Defined Networking (SDN). The concept behind SDN is to allow centralized control of network traffic to provide end-to-end services at an application layer. WLAN administrators will see similarities with various AP to controller-tunneling protocols such as CAPWAP and LWAP. With SDN networks, administrators will be able to leverage the application's communications with the network management system to provide network virtualization and automation of configuration across the entire network not just the WLAN. This will allow administrators to support an application's performance and security requirements independent of the client device. To accomplish this layer 7 granularity of network traffic is required. Application vendor's support with APIs will make it easier to provision. Communications vendors are already beginning to develop support. SDN interoperability will most likely take a considerable amount of time, as there is no defined standard. Networking vendors that make both wired and wireless components will be first to offer the complete end-to-end solution. Environments with mixed infrastructure will have challenges supporting SDN application flows until standards are developed.

IPv6   IPv4, an IP addressing system, was developed in 1981 and was created with a known limitation of 4 billion public IP addresses. These were believed to be sufficient for any foreseeable growth. With the explosive growth of mobile devices globally, it became clear very quickly that an updated addressing system is required. IPv6 was created in the mid-1990s to address the lack of scalability of IPv4. Some IPv6 adoption statistics presented by Dr. Geoff Huston at the ISOC INET conference, just prior to the summit itself, are outlined below.

- It's taken the United States 2 years of geometric growth to get IPv6 to 2.5 percent as a percentage of overall Internet traffic.
- 1.2 percent of users accessed Google over IPv6.
- Average usage of IPv6 across the Internet is at 1.2 percent.
- 50 percent of the world's transit ISPs support IPv6.
- Meanwhile, 56 percent of ISPs in North America don't support IPv6.
- 50 percent of the hosts connected to the Internet have support for IPv6.

Given the statistics above, wide-scale adoption of IPv6 is more than 2 to 3 years away. Network address translation (NAT) which can have some major ramifications on PTP voice and video applications from devices outside the hospital network to devices within the network.

In addition IPv6 brings more classes of service, and thus potentially higher QoS. This can have a significant impact on eHealth, mHealth, and telemedicine applications.

802.11u/Hotspot 2.0/Passpoint   The newest emerging technology for guest Wi-Fi goes by several names depending on who you ask. For the IEEE it might be 802.11u, for the Wi-Fi Alliance (WFA) it would be Hotspot 2.0 Passpoint, or next generation hotspot (NGH) from the Wireless Broadband Alliance (WBA). The purpose of these new technologies is primarily to offload cellular (3g/4g/LTE) to Wi-Fi and add more security as a secondary goal. For the remainder of the chapter we will refer to this collection of technologies as Hotspot 2.0, which was the task group of the Wi-Fi Alliance. Hotspot 2.0 must support EAP-SIM. The SIM card in a phone is used to authenticate a phone or tablet to a Wi-Fi network just like it would be authenticated to a cellular network. This can also provide seamless roaming between cellular and Wi-Fi networks. EAP-SIM will working with GSM networks (think AT&T and T-Mobile in the United States). Meanwhile, EAP-AKA would likely be used by carriers like Verizon and Sprint as it is built into a UMTS Subscriber Identity Module (USIM). It should be noted that both the client device and Wi-Fi infrastructure must support Hotspot 2.0. This will certainly dictate a slow adoption rate as both will need to be updated to add this functionality. You may have already experienced EAP-SIM authentication with a Hotspot 2.0 network. AT&T has begun to integrate this into its phones and business partner Wi-Fi networks. The next time you are at a popular fast-food chain, you may notice that you are already connected to their Wi-Fi even though you did not specify to do so. That is the beauty of a technology like this. Earlier in the chapter we discussed balancing ease of use with anonymity. In this scenario, the guest is not anonymous but is able to be connected without even knowing. From a supportability standpoint this is the Holy Grail of guest Wi-Fi registration. They are registered with the WLAN without doing anything. In addition, this wireless network can be encrypted. While this is mostly being used by commercial telecommunications partners to provide defined hotspots for customers, there is a tremendous opportunity in enterprise spaces such as healthcare.

mHealth

The field of mHealth has seen a tremendous amount of growth over the last several years. It has been expanding so aggressively that the HIMSS organization has developed an arm mHIMSS in 2012 dedicated to mobility, and the mHealth field. This field continues to shape the evolution of patient care. The use cases range from smaller portable medical devices to sophisticated medical applications, and an ever-increasing market for home health. The field is vast and the type of mobility that it leverages is beyond Wi-Fi. With its bandwidth, and heavy adoption, Wi-Fi has and will continue to play a vital role in this space. To date we have seen medical apparatus integrated to smart phones and tablets, which then transmit the data to a central repository or directly to a clinician. In addition, applications oriented at customer engagement and wellness are becoming increasingly popular. Many view the focus on wellness as a step in the right direction to decreasing the costs of patient care. Ensuring that the Wi-Fi networks in hospitals can intelligently secure and allocate bandwidth to mission critical applications will help ensure that we are ready for these emerging applications.

Beyond these trends and what seems like an addiction to smart phones and tablets, the next form factors are exciting. We are still tethered to our devices. If that tether can be removed, productivity can be increased exponentially, but we have yet to see a creative form factor, beyond Google glasses, that can accomplish this. The future of Wi-Fi in healthcare is an exciting one full of opportunities and challenges.

Index